Eustasy, High-Frequency Sea-Level Cycles and Habitat Heterogeneity

Eustasy, High-Frequency Sea-Level Cycles and Habitat Heterogeneity

Mu. Ramkumar
Periyar University, Salem, Tamil Nadu, India

David Menier
Université Bretagne Sud, Vannes, France

ELSEVIER   elsevier.com

Elsevier
Radarweg 29, PO Box 211, 1000 AE Amsterdam, Netherlands
The Boulevard, Langford Lane, Kidlington, Oxford OX5 1GB, United Kingdom
50 Hampshire Street, 5th Floor, Cambridge, MA 02139, United States

Notices
Knowledge and best practice in this field are constantly changing. As new research and
experience broaden our understanding, changes in research methods, professional practices, or
medical treatment may become necessary.

Practitioners and researchers must always rely on their own experience and knowledge in
evaluating and using any information, methods, compounds, or experiments described herein. In
using such information or methods they should be mindful of their own safety and the safety of
others, including parties for whom they have a professional responsibility.

To the fullest extent of the law, neither the Publisher nor the authors, contributors, or editors,
assume any liability for any injury and/or damage to persons or property as a matter of products
liability, negligence or otherwise, or from any use or operation of any methods, products,
instructions, or ideas contained in the material herein.

British Library Cataloguing-in-Publication Data
A catalogue record for this book is available from the British Library

Library of Congress Cataloging-in-Publication Data
A catalog record for this book is available from the Library of Congress

ISBN: 978-0-12-812720-9

For Information on all Elsevier publications visit our
website at https://www.elsevier.com/books-and-journals

www.elsevier.com • www.bookaid.org

Publisher: Candice G. Janco
Acquisition Editor: Louisa Hutchins
Editorial Project Manager: Tasha Frank
Production Project Manager: Stalin Viswanathan

Typeset by MPS Limited, Chennai, India

CONTENTS

PREFACE

The present environments and ecosystems are an evolving continuum with those of the past and the future. Estimations on geological–biological dynamics through the study of habitat heterogeneity and biotic abundance are essential for understanding the continuum. The earth processes operate at various timescales to create unique hierarchical sedimentary facies succession under the dominant influence of global sea-level cycles, in which, the other controls, namely, tectonics, climate, and sediment influx either get imprinted or supercede at times of favorable milieu. Nevertheless, the sea-level fluctuations are responsible for the dynamics of ecological niches especially in the marine realm. The sea-level changes (either eustatic or influenced by other phenomena) might also alter the oceanographic phenomena, such as oxygen depletion and water-mass stratification, which in turn control biodiversity. Thus, habitat heterogeneity, as explicit in terms of spatial–temporal variations in paleogeomorphic features and facies types, occurrence, absence, diversity, and density of taxa, needs to be studied in the context of eustatic sea-level changes, perturbations, etc.

Though studies abound on paleontological database, the impacts of dynamics of tectonic-climatic-sea-level-sediment influx on facies sequences, species diversity, proliferation, and habitat require attention. Further, a disparity of understanding on the term habitat in geological perspective exists. Few specialized articles are available; but the lacuna is wide and diverse and needs to be filled by a bridge-level reference material. In this book, we attempt defining the habitat, provide a start-up level classification of habitat, and demonstrate a case study for recognizing habitats in geological records through proxies. We have demonstrated the nexus between tectonics-climate-sea-level fluctuations-sediment influx/production and the facies pattern, which in turn influenced the area of ecological niches and thereby the biotic occurrence and diversity. It is our endeavor to encourage the beginners to conduct research in this interdisciplinary field of geosciences.

Mu. Ramkumar and David Menier

ACKNOWLEDGMENTS

Nomination of MR as National Working Group Member to IGCP-609 Cretaceous Sea-level Changes has rekindled the interest on the sea-level changes and resultant impacts on depositional environment, biodiversity, and habitat. The interest grew into penning a trilogy of books documenting nature and characteristics of sea-level fluctuations, biodiversity trends, and habitat dynamics. This book, third of this trilogy, traces the importance of understanding habitat heterogeneity in the context of spatial and temporal scales of facies changes. Seeds for working on this book were sown during reading the works on gigantism of Paleozoic invertebrates by C. Klug, K. de Baets, B. Kröger, M.A. Bell, D. Korn, and J.L. Payne, and Eocene-Miocene biodiversity of Sarawak carbonates by M. Mihaljević, W. Renema, K. Welsh, and J.M. Pandolfi and by the suggestion of C. Klug to analyze the pattern of environmental and climatic influences on species diversity and population. C. Klug, Institüt für Paläontologie und Paläontologisches Museum, University of Zurich, is thanked for suggesting MR to ponder over the issue with reference to the Cauvery Basin. Research on the Cauvery Basin was initiated with the financial grants from Alexander von Humboldt Foundation, Germany. Prof. Dr. Doris Stüben, Institute of Mineralogy and Geochemistry, erstwhile University of Karlsruhe, currently rechristened to Karlsruhe Institute of Technology, Germany is thanked for her invitation to MR to conduct collaborative research. MR learnt a lot through intense and lively discussions with Dr. Zsolt Berner, Senior Scientist and Head of Laboratories, Institute of Mineralogy and Geochemistry, Karlsruhe Institute of Technology (erstwhile University of Karlsruhe), Germany. Partial funding to this work in the form of Research Associateship and Senior Research Associateship was provided to MR by Council of Scientific and Industrial Research, New Delhi, and University Grants Commission, New Delhi in the form of research project (Maastrichtian-Danian part of the study area). The authors thank the scientific personnel of Institute namely, Dr. Utz Kramar for XRF analytical facilities, Dr. Karotke and Mrs. Oetzel for XRD analyses, Mr. Predrag Zrinjsak for carbon and sulfur analyses and Dr. G. Ott for computing facilities. Shri. T. Sreekumar, Geologist, OFI, Mumbai, is thanked for assistance during the field survey. MR thanks his wife A. Shanthy, daughter Ra. Krushnakeerthana, and son Ra. Shreelakshminarasimhan for their understanding while authoring this book. DM thanks his wife Nathalie and children Baptiste, Louis, and Périnne for their active support.

DM and MR thank Dr. Manoj Mathew, UBS, Laboratoire Geosciences Ocean, UMR CNRS 6538 for helps in editing the figures. Dr. Sandrine Maximillien, Service for Science and Technology, French Embassy in India, Mumbai, and "Laboratoire d'Excellence" Labex MER (ANR-10-LABX-19) (Axis 4: Sediment transfers from the coast to the abyss) and cofunded by a grant from the French Government under the program "Investissements d'Avenir" are acknowledged for extending financial and other support for the visits of the authors to each other's institutions in India and in France.

The staff of editorial, back office, and production teams of Elsevier extended constant encouragements. Candice G. Janco, Louisa Hutchins, Tasha Frank, Stalin Viswanathan and Sandhya Narayanan deserve special mention and are acknowledged for their professional handling and postproduction support. MR is thankful to The Lord Shree Ranganayagi Samedha Shree Ranganatha for HIS boundless mercy showered and by whose ordinance those acknowledged in this section have shouldered the responsibilities either actively or passively.

Mu. Ramkumar and David Menier

CHAPTER 1

Defining Habitat and Habitat Heterogeneity

1.1 INTRODUCTION

The present environments and ecosystems are an evolving continuum with those of the past and the future. Understanding this continuum necessitates estimations on geological—biological dynamics through documenting the habitat heterogeneity and biotic abundance (Fioroni et al., 2015) of the present and the past. Comparative analyses of both living and fossil biota are necessary to understand biodiversity and evolutionary patterns of the Earth's history (Roy et al., 1998). Chen et al. (2014) noted that climate extremes and associated environmental stresses may become a characteristic of our future. Modeling future biodiversity losses and biosphere—climate/environment feedbacks are inherently difficult. An important tool to address this question is to document and understand similar events of the geologic past where magnitude and/or rates of change in the global climate/environment system destabilized the biodiversity, leading at times to mass extinctions. Coordinated study on the responses of habitats and biota thrive in them to environmental parameters associated with global change has ecological and geological implications (Hallock, 2005; Crampton et al., 2011; Sessa et al., 2012). Documenting habitat diversity patterns through geological past is key for understanding the significance of biodiversity origins and patterns through time (Mihaljević et al., 2014). And, this is the phenomenon that forms focal theme in this book, especially the habitats and biotic changes of the marine ecosystems.

1.2 DEFINITION OF HABITAT HETEROGENEITY

The Merriam Webster© dictionary defines "habitat" as "the place or type of place where a plant or animal naturally or normally lives or grows." Wikipedia© defines habitat as a "zone in which an organism lives and where it can find food, shelter, protection and mates for reproduction." Published geological literature portrays highly differing

Eustasy, High-Frequency Sea-Level Cycles and Habitat Heterogeneity.
DOI: http://dx.doi.org/10.1016/B978-0-12-812720-9.00001-2

perspectives on this term. It is used interchangeably with many terms, including ecosystem, niche, and substrate conditions, and others. Various parameters are also in use for defining habitats. For example, altitude (Cantalapiedra et al., 2012; Sammarco et al., 2016), bathymetry (Flessa et al., 1993; Bonis et al., 2010; Holland, 2012; Mihaljević et al., 2014; Brom et al., 2015; Johnson et al., 2015; McClure and Rowan Lockwood, 2015; Reich et al., 2015; Riegel et al., 2015; Baucon and Neto de Carvalho, 2016; Kröger et al., 2016), climate zone delimited by latitudes (Cook et al., 2014; Hoffmann et al., 2014; Novak and Renema, 2015; Reich et al., 2015), lifestyle of organisms such as benthic, planktonic, and others (Bambach, 1977), marine (Kalmar and Currie, 2010; Mihaljević et al., 2014; Mancosu and Nebelsick, 2015; Tanabe et al., 2015; Ungureanu et al., 2015; Novack-Gottshall, 2016; Wignall et al., 2016) and nonmarine/continental regions (Kalmar and Currie, 2010; Kennedy and Droser, 2011; Neto et al., 2014), photic conditions (Sammarco et al., 2016), circulation/open-closed conditions of marine region (Johnson et al., 2015; Retallack, 2015), turbulence (Baucon and Neto de Carvalho, 2016), turbidity (Santodomingo et al., 2015), salinity of aquatic ecosystems such as fresh, brackish, saline, and others (Jarzen and Klug, 2010; Kalmar and Currie, 2010; Romano and Whyte, 2013; Cook et al., 2014; Suarez-Gonzalez et al., 2015), vegetated and nonvege-tated nature (Villafaña and Rivadeneira, 2013; Albano, 2014; Reich, 2014; Steffen, 2016), natural forests and cultivated/plantation region (Ricketts et al., 2015), area inhabited by specific taxa/organism (Reymond et al., 2011; Darroch, 2012; Bracchi et al., 2015; Richey and Sachs, 2016), or specific function of organisms such as nesting (Smith and Hayward, 2010) and nature (firm or soft or sandy, etc.), or compo-sition (siliciclastic, carbonate, mixed siliciclastic, etc.) of substrate (Holland and Christie, 2013; Sessa et al., 2013; Archuby et al., 2015) are used for defining/delimiting/characterizing/distinguishing habitats.

Review of previous publications and on a perspective of practicality in geological context showed that the term "habitat" can be defined by an area, either terrestrial, aquatic, or otherwise that provides ecosystem services for biota to engage in life functions. There are documented evidences of life functions of organisms that create and modify habitats (Perry et al., 2015). Recently, Novack-Gottshall (2016) defined it as "functional-trait space." As explained in the following section and other chapters, "habitat" of organisms should be defined by a defini-tive physical, chemical, and biological attributes on a geographic domain, characterized by specific attitudinal and other parameters. In this regard, the definition of "ecological niche" as "geographically explicit information on species occurrences and the suites of environmental conditions experienced at each occurrence point" (*sensu stricto—* Myers et al., 2015) seems to be more appropriate and synonymous with "habitat" in geological connotation. More specifically, the consideration of subenvironments, recognizable in terms of distinct facies, geomorphic, and species accumulation as synonymous with distinct "habitat" attempted by many authors including, but not limited to, Reymond et al. (2011), Cantalapiedra et al. (2012), Nürnberg and Martin Aberhan (2013), Archuby et al. (2015), Diaz-Martinez et al. (2015), Wilson (2015), and Ritterbush et al. (2016) is appropriate.

The term "habitat heterogeneity" is defined by the variation of the area in spatial and temporal scale. The factors that significantly influence the habitat heterogeneity are those espoused for sequence development, namely tectonics, climate, sea-level fluctuation, and sediment influx/production. Among these factors, the sea-level fluctuations or relative sea-level cycles play vital and direct role in biotic occurrence, extinction, population, and diversity, which in turn reflect as a function of habitat heterogeneity.

1.3 DEFINITION OF "SPACE" IN HABITAT

Biodiversity hotspots are geographical regions characterized by exceptionally high species diversity (Reid, 1998). The diversity is facilitated through diverse habitats (MacArthur and MacArthur, 1961; Sessa et al., 2012) and conducive environmental factors (Ramkumar, 2015a) for biotic proliferation (Ramkumar, 2015b). Given cognizance to these together with the definition of "habitat" presented in the previous section, geological characterization of "extent" or "space" of unique habitat requires defining "areal distribution" over a recognizable "time duration." While the time duration of prevalence of specific habitat can be defined using the geochronological chart, the "space" requires specifications. The geological literature shows examples of defining extent of habitat as large as latitudinal climate zone to as small as specific functional area of an organism. For example, Valentine and Jablonski (2010) demonstrated the existence of a latitudinal diversity gradient in bivalves through time. It suggested variance of habitat on a latitudinal scale. The studies of Roberts and Ormond (1987) and

Mihaljević et al. (2014) have shown the delimitation of habitats at microenvironmental scales. According to these authors, in shallow-marine environments (especially shoals, inner-ramp, and platform margin), habitat heterogeneity is typically facilitated through a range of tidal regimes and wave energy. Considering these two extremes (latitudinal gradient and limited geographic area as defined by geomorphic attribute), and the definition of sedimentary environment (*sensu* Reineck and Singh, 1980), the sprawl of specific geomorphic unit can be assigned to delimit space for habitats. This is hierarchical, ranging from macro-, meso-, and microgeomorphic units. Nevertheless, the extent of habitat is relative and thus delimited by the context and objective of discussion/interest.

Given cognizance to these definitions, the habitats in the geological context can be defined/recognized in terms of geomorphic expressions and their subdivisions ranging from as large as climatic zones (delimited by latitudinal boundaries) to as small as tidal zones or further smaller geomorphic units. For example, as depicted in Fig. 1.1, an

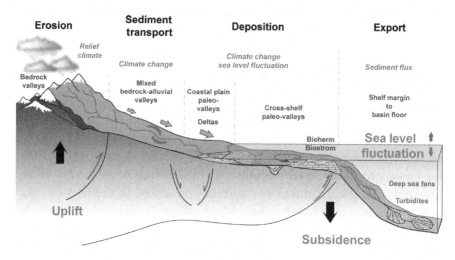

Figure 1.1 Idealistic profile of source-sink tectonogeomorphology. The figure shows the generalized regional geomorphic profile across mountain-ocean basin. The four major factors that are considered as basic tenants of sequence development namely, tectonics-climate-relative sea-level-sediment influx/production that create and modify the terrestrial−marine habitats and resultant changes/shift are also indicated in the figure. For example, any change in the bathymetry might impact/shift/destruct/expand the locales of carbonate production. Similarly, tectonic and climatic events control the weathering in the continental regions, sediment production/influx, and impact the terrestrial−marine habitats. Thus, the factors controlling sequence development need to be considered for analyzing the habitat dynamics. (Inspired from Blum, M., Martin, J., Milliken, K. and Garvin, M., 2013. Paleovalley systems: insights from Quaternary analogs and experiments. Earth Sci. Rev. 116, 128−169)

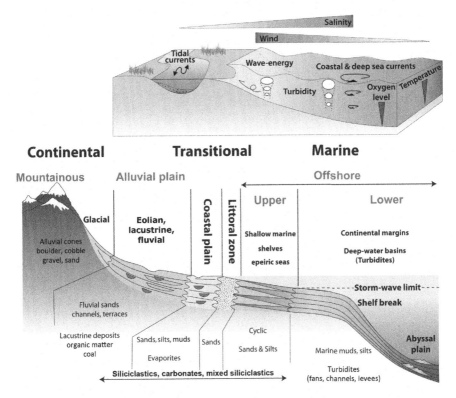

*Figure 1.2 **Depositional systems with diverse substrate and environmental parameters.** These are the large-scale depositional systems that may be subjected to changes due to fluctuations in sea-level. Associated with the changes in depositional systems are their shift/expansion/destruction which in turn change substrate conditions and environmental parameters such as energy, turbidity, salinity, oxygen, and others. All these impact the biotic proliferation, diversification and dwindling/extinction patterns. The loci of individual habitats, thus, might be in a dynamic equilibrium and may adjust to newer equilibrium as the major governing factors change.*

idealistic profile of geomorphic expressions from highest to lowest elevations represent a wide variety of sedimentary systems, covering comprehensive source-sink (*sensu* Balseiro et al., 2011) habitats (Blum et al., 2013). Each of the sedimentary system, as depicted in Fig. 1.2, represents a finite range of geomorphic units, recognizable through a variety of attributes detailed in successive chapters in this book.

CHAPTER 2

Sequences, Sea-Level Cycles, and Habitat Dynamics

2.1 SEQUENCES AND HABITATS

The basic consideration in the theory of sequence development is that, the oscillations in sea-level significantly shift the principal loci of deposition and erosion. This, in effect, changes the area of habitats and areal distribution of habitats as well (Bonis et al., 2010; Holland, 2012; Holland and Christie, 2013; Nürnberg and Aberhan, 2013; Wignall et al., 2016). When sea-level falls, the shoreline shifts to former offshore, and the areas previously occupied by marine ecosystem is exposed subaerially allowing the terrestrial and fluvial ecosystems to establish themselves there. Similarly, when sea-level rises, it shifts the shoreline over continental regions, resulting their submergence and the terrestrial and fluvial ecosystems are occupied by the coastal and holomarine ecosystems (Smith et al., 2006; Ruban, 2009, 2010a,b; Rook et al., 2013). However, few other authors including Holland (2012), based on the measurements on change in shallow marine regions from data bases of global elevation, opined that in contrast to prevailing views, sea-level rise does not consistently generate an increase in shelf area, nor does sea-level fall consistently reduce shelf area. Accordingly, different depth-defined habitats on the same margin with different changes in area for the same sea-level change and different margins with different changes in area for the same sea-level change were also reported to occur. The extent of area affected by these changes depends on various factors (Smith and Benson, 2013) including the starting position, magnitude, and direction of sea-level change (Holland, 2012), slope, climate, relative contribution by tectonism, and others. In the context of habitat dynamics, the shifts of areal extents of area submerged and exposed and relict structure (Menier et al., 2014a, 2016; Pian et al., 2014), and others, as the case may be, influence the change in habitat and the corresponding paleobiodiversity trends (Smith and Benson, 2013).

Eustasy, High-Frequency Sea-Level Cycles and Habitat Heterogeneity.
DOI: http://dx.doi.org/10.1016/B978-0-12-812720-9.00002-4

The transgressions and regressions might have connected and disconnected the ancient seas and oceans and permitted and restricted the migration of biota, respectively. These might have also strengthened/diminished the occurrence and abundance of biota. Thus, the occurrence, absence, population, and diversity of ancient biota might have been under the control of oscillations of sea-level. Though consensus exists on the impact of sea-level fluctuations over paleobiodiversity trends, the lack of perfect synchroneity among them was reported to be the result of diverse response of depth−area relationships of habitats with reference to a given magnitude of sea-level fluctuation. In a classic study, Holland and Christie (2013) demonstrated highly specific response of habitat area to sea-level change. According to these authors, a particular bathymetric zone on a particular margin may vary substantially in area during a sea-level change, while other such zones and margins may experience little or even opposite responses. Through construction of idealistic models for shallow, moderate, and deep water habitats, these authors have reported that the intermediate-depth and deep-water regions with sinusoidal variations in width, while the shallow water region showed nonsinusoidal change of width. Similarly, the shallow water region showed abrupt widening at the beginning of the highstand systems tract and also abrupt narrowing at the onset of sea-level fall at the beginning of the falling-stage systems tract. The study of Tibert and Leckie (2004) is a case in point where distinct vertical partitioning of taxon associations between high and low marsh, i.e., at an altitudinal change of 1 m was demonstrated. It also emphasized the magnitude of impact the sea-level fluctuations (however minor may be) over the abundance and diversity of biota. Though this premise is implicit in the sequence and sea-level research, studies documenting the relationships between the relative sea-level fluctuations and the resultant habitat heterogeneity and attendant species diversity, population dynamics, and others are few. Though few excellent publications have documented the relationships between the sea-level fluctuations, especially the eustatic cycles and biodiversity (Smith and Benson, 2013), they were specific to selective fauna (O'Dogherty et al., 2000; Sandoval et al., 2001a,b; Yacobucci, 2005; Purdy, 2008) or flora (Martin et al., 2008) or restricted to certain chronologic intervals (Hallam and Wignall, 1999; Gale et al., 2000) that limited the full appreciation of the link between the habitat heterogeneity (Holland, 2012), paleobiodiversity and the sea-level fluctuations.

Changes in sea-level alter the area of epicontinental seas and determine the area available for both biotic habitats (Simberloff, 1974; Crampton et al., 2011; McClure and Rowan Lockwood, 2015; Wignall et al., 2016) and sediment accommodation in similar ways (Newell, 1952; Simberloff, 1974; Crampton et al., 2011). Changes in the physical environment are major drivers of evolutionary change, either through direct effects on the distribution and abundance of species or more subtle shifts in the outcome of biological interactions (Leonard-Pingel and Jackson, 2016). Habitat heterogeneity over geographic space is a strong determinant of taxonomic richness (Rook et al., 2013). Smith et al. (2006) suggested examining biodiversity in response to sea-level fluctuations such that deeper-water species should appear first in offshore settings and migrate into more onshore settings when sea levels have risen. During sea-level lowering and sea-level lowstand, migration of shallow-water taxa into shelf might be expected. This suggestion stands confirmed from the study of Powers and Bottjer (2009) who have documented elevated bryozoan extinction rates during the Late Permian and Late Triassic that were coupled with major changes in their habitats. They reported a permanent change in the paleoenvironmental preferences of bryozoans from nearshore to mid-shelf settings.

In one of the systematic documentation and analysis of biodiversity as a function of habitat dynamics influenced by sea-level cycles, Smith et al. (2006) recorded about 3500 occurrences of echinoid and correlated against sequence stratigraphy representing onshore, mid-shelf, and deeper-shelf habitats. According to these authors, the ranges of individual taxa expand and contract across the shelf as sea levels change. In mid-shelf environments, more onshore taxa appeared only near sequence bases at times of lowest sea-level, while those from more outer shelf settings are found during highstand intervals, and this created a cyclic pattern of diversity. A link between habitat heterogeneity and taxonomic diversity is another way in which the geologic processes might directly influence living organisms to yield correlated patterns in the rock and fossil records. Peters (2008) demonstrated the relationship between Phanerozoic marine biodiversity and the temporal and spatial distributions of carbonate and siliciclastic lithofacies. As sequences, parasequences and facies bundles (*sensu* Catuneanu, 2006; Catuneanu et al., 2011) are the products of unique environmental parameters, introduced to the depocenters by relative sea levels and other controlling factors of sequence development, the distribution, stacking pattern, and contact relationships of unique facies types, created in a specific geomorphic set up, might possess unique environmental conditions, and in turn exerted influence over the temporal and spatial biodiversity and habitat trends. Many studies including Klug et al. (2006), Holland and Zaffos (2011), McMullen et al. (2014), and Villafaña and Rivadeneira (2014), among others, have either upheld or documented this relationship between unique habitats—facies patterns and associated unique paleobiodiversity patterns. Zuschin et al. (2014) are of the opinion that understanding their ecological dynamics over geological time scales requires compilation of paleontological data in a sequence stratigraphic framework, which in turn allows evaluation of paleocommunity dynamics in an environmental context. McMullen et al. (2014) provided a short review on previous studies documenting the relationship between fossil occurrences and the sequence architecture. These authors reported strong facies control and fossil concentrations at maximum flooding surfaces, in the upper portion of parasequences, and within lags overlying sequence boundaries. Based on their review and study, these authors have concluded that sequence stratigraphic architecture may be a useful approach for discovery of marine vertebrates and that sequence stratigraphic context should be considered when making paleobiological interpretations of marine vertebrates as well as invertebrates.

2.2 SEA-LEVEL FLUCTUATIONS AND ENVIRONMENTAL PARAMETERS

The biota (either terrestrial or marine) depend on the environmental conditions for survival, proliferation, and diversification. On a geochronological scale, the changing climatic, tectonic, and other conditions might have influenced the distribution of suitable/adverse conditions of thriving of biota. This influence has a long-term implication as well, as stated by Leonard-Pingel and Jackson (2016), that changes in the physical environment are major drivers of evolutionary change, either through direct effects on the distribution and abundance of species or more subtle shifts in the outcome of biological interactions. On a long-term trend, as Ruban (2010a) and few other authors (Simberloff, 1974; Crampton et al., 2011; Holland, 2012; Rook et al., 2013) demonstrated, the sea-level fluctuations were responsible for the dynamics of ecological niches. For example, the estuarine, coastal, and shallow marine regions, particularly located within the photic zone support extensive primary production and serve as habitats to primary, secondary, and tertiary consumers of epi and infaunal benthic life as well as nektobenthics and nektons. There are documented evidences of habitat tracking nature of taxa (Zuschin et al., 2014) during geologic past which implies environmental preferences of organisms (Wilson, 2015). Hence, the relative sea-level fluctuations might affect the habitats, environmental niches (Holland and Zaffos, 2011), and thereby the occurrence, distribution, population, and diversity (Sessa et al., 2012) of the lives that thrive in the zone of sea-level fluctuation. Novack-Gottshall (2016) documented evolutionary dynamics of paleotaxa as a function of ecospace (habitat) diversification. Linear models of paleo-biodiversity trends showed significant independent effect of geographic range and habitat breadth on genus durations, diversity dynamics, extinction risk, and origination (Nürnberg and Martin Aberhan, 2013).

Geographic variations in diversity are associated with tropic resources and planktonic modes (Benton, 2013). Nevertheless, influence of other environmental factors such as temperature, turbidity, depth, substrate, hydrodynamic energy, and pH was observed to have exercised to have control (e.g., Hottinger, 1983; Connell, 1997; Renema and Troelstra, 2001; Ando et al., 2010; Crampton et al., 2011; Zuschin et al., 2014; Wilson, 2015; Sælen et al., 2016). These environmental factors are under the control of variations in regional tectonics, sea-level, climate, and oceanographic parameters (Rosen, 1984) and ultimately reflect in the habitat and biodiversity patterns (Faith and Behrensmever, 2013). Observations on the recent ocean warming, acidification, and deoxygenation revealed a dramatic effect on the flora and fauna of the oceans with significant changes in distribution of populations and decline of sensitive species (Bijma et al., 2013). According to MacLeod (1994), moderate faunal homogeneity of Cretaceous habitats was the result of low thermal gradient of sea surface temperatures. As the sea surface temperature might have been under the direct influence of atmospheric $p\mathrm{CO}_2$ and temperature, and result in glacial advancement/retreat followed by sea-level retreat/advancement, respectively, elevated $p\mathrm{CO}_2$ in the atmosphere and resultant greenhouse effect warms the oceans, initiates destabilization of ecosystems, reduces the species populations, and limits the potential for adaptation (Bijma et al., 2013). It also leads to shifting of species ranges poleward (Clapham and James, 2012) and into deeper, cooler waters. The habitat shifts within short timeframes may lead to increased extinction risk for long-lived, slow growing, sessile habitat-forming species. In the case of coastal species, a poleward shift in distribution may be limited by geography as organisms simply *run out* of coastline to migrate (Bijma et al., 2013).

In addition to the change in the area under sea, the sea-level changes might also alter the oceanographic phenomena, such as oxygen depletion and water–mass stratification, which in turn exercise important controls on biodiversity. Retallack (2011) and Sigwart et al. (2014) listed rapid burial (obrution), stagnation (eutrophic anoxia), fecal pollution (septic anoxia), bacterial sealing (microbial death masks), brine pickling (salinization), mineral infiltration (permineralization and nodule formation by authigenic cementation), incomplete combustion (charcoalification), desiccation (mummification), and freezing as the criteria/events that support exceptional preservation of paleotaxa in abundance. This list also exemplifies the multitude of environmental parameters and geological processes associated/controlled by those parameters affecting biodiversity distribution and trends. However, exceptions occur, such as the larger benthic foraminifera that show tolerance to environmental conditions and occur in shallow, tropical marine habitats, and mixed carbonate–siliciclastic depositional settings (Novak and Renema, 2015). On the contrary, Zuschin et al. (2014) recorded a gradual transition from depositional bathymetry and hydrodynamic energy that resulted in similar gradual transition in biotic response, affirming the dependence of biodiversity trends over environmental conditions. The gradient of environmental parameters was found to be the strong determinant of diversity patterns (Hautmann, 2014).

2.3 ENVIRONMENTAL PERTURBATIONS AND HABITAT CHANGES

Luxuriant environmental condition results in proliferation and diversification while the environmental stress often results in dwindling (Suarez-Gonzalez et al., 2015) and extinction. Clapham and James (2012) noted that marine invertebrates face extinction when climate changes outpace their ability to adapt to the stress. Based on worldwide search, Retallack (2011) also documented that times of especially widespread exceptional fossil preservation were also times of stage boundaries, mass extinctions, oceanic anoxic events, carbon isotope anomalies, spikes of high atmospheric CO_2, and transient warm—wet paleoclimates in arid lands. Cherns et al. (2008) opined that if the initial microstructural and mineralogical characteristics of certain organisms such as molluscs that preferentially get destroyed and their fossilization potential is less than many of the contemporary organisms and skew the biodiversity computations, the use of storm beds, shell plasters and submarine hardgrounds, and low energy, organic-rich mud-dominated settings, can be identified as fossil deposits that can preserve the labile aragonitic component of the fauna and thus represent potential taphonomic windows. However, as explained in following chapters, the creation of taphonomic windows and associated environmental parameters are also under the influence of sea-level fluctuations. For example, Cherns et al. (2008) commented that sedimentation by storm event concentrates epifaunal shells and shallow infauna exhumed by scour, in rapidly buried deposits. Shell beds resulting from single event as well as composite (i.e., time-averaged) processes can be recognized as proxies for the dominant faunal components.

Gradual decrease in gastropod size through the Guadalupian—Lopingian interval (Payne, 2005), culminating in the predominance of small individuals in the Permian—Triassic extinction interval and the Early Triassic, was recorded (Fraiser and Bottjer, 2004; Brayard et al., 2010). The long-term size diminution may have resulted from increasing but low levels of environmental stress, potentially a precursor to the environmental crisis that caused the end-Permian mass extinction and prolonged Early Triassic recovery (Clapham and Bottjer, 2007). Powers and Bottjer (2009) recorded environmental stress and resultant changes in diversity, ecology, and paleogeography of Middle-Phanerozoic bryozoa. Similar observations could be made from the Cauvery Basin with reference to the *exogyra* (<1 cm in size in Karai Formation and few cm in Kallankurichchi Formation), *rhynconella* (<1 cm in Garudamangalam Formation). Conversely, the gigantism of *gryphea*, *terebratula*, and *stigmatophygus* found in the Kallankurichchi Formation, Cauvery Basin, India, indicates luxurious environmental conditions during Early Maastrichtian in this basin (Ramkumar and Chandrasekaran, 1996).

CHAPTER 3

Influence of Sea-Level on Facies and Habitat Heterogeneity

3.1 DEPOSITIONAL SYSTEMS AS DEFINED BY END-MEMBERS

Sedimentary environments (*sensu* Reineck and Singh, 1980), or subenvironments (*sensu* Ramkumar, 2001; Wilson, 2015), or depositional setting (*sensu* Mihaljević et al., 2014) or depositional systems (*sensu* Ritterbush et al., 2014; Suarez-Gonzalez et al., 2015) are unique areas with definable areal extent, geographic distribution, and geomorphic attributes (Cantalapiedra et al., 2012; Bracchi et al., 2015; Perry et al., 2015; Sammarco et al., 2016), in which unique set of physical, chemical, and biological conditions prevail. Published literature abounds with recognition/characterization of unique habitats in terms of mappable geomorphic units (e.g., Ritterbush et al., 2014; Bracchi et al., 2015; Mancosu and Nebelsick, 2015; Perry et al., 2015; Zuschin and Ebner, 2015). It leads to the surmise that prelevance of unique environmental conditions at a given place for certain duration results in unique facies types and or a unique combination of facies associations in stratigraphic (temporal) scale (Smith and Benson, 2013). Precise characterization methods and classification systems to recognize and distinguish unique facies and facies associations are available. The durations of temporal scales are measured in relative as well as absolute age data. The introduction of sequence stratigraphic concepts and the relative sea-level cycles of long term as well as up to millennial durations enabled typification of unique facies types and facies associations distinguishable from adjacently located (spatial and temporal) units. Thus, the depositional systems, each recognized at required spatial and temporal scale might represent unique habitat types. The spatial and temporal distribution and diversity of the depositional systems might reflect the diversity of habitats. In order to recognize habitats from geological records, the tools available for characterization and delineation of individual depositional units, ascribable to lithological, paleontological, bulk compositional, and other properties can be used.

Eustasy, High-Frequency Sea-Level Cycles and Habitat Heterogeneity.
DOI: http://dx.doi.org/10.1016/B978-0-12-812720-9.00003-6

Thus, the end-members of classificatory systems of rock records, such as litho, bio, seismic, and chemo facies types can be utilized to identify unique habitats.

Published literature depicts recognition of independent habitat in terms of unique geomorphic identity and thus with unique set of environmental or depositional conditions and biotic composition, distribution, and others. For example, Ritterbush et al. (2014) recognized unique habitats in tune with midshelf, carbonate ramp settings, and others. Zuschin and Ebner (2015) recognized the biodiversity differences between habitats and within habitat and reasoned to be the result of population dynamics recorded from intertidal, subtidal, and lagoonal regions. Brady (2016) recognized within habitat (shallow to deep subtidal region) changes in skeletal composition. According to Badgley and Finarelli (2013), modern ecosystems, especially the regions of topographic heterogeneity, support high species densities, and this biogeographic pattern is due to either greater diversification rates or greater accommodation of species in topographically complex regions. Applying this premise over Neogene records of rodent diversity for three regions in North America, these authors concluded that climatic changes interacting with increasing topographic complexity intensify macroevolutionary processes. In addition, close tracking of diversity and fossil productivity with the stratigraphic record suggested either large-scale sampling biases or the mutual response of diversity and depositional processes to changes in landscape history. These interpretations exemplify the roles of tectonics, climate, and geomorphology over habitat and biodiversity patterns.

3.2 SCALES OF DEPOSITIONAL SYSTEMS

A variety of depositional systems can be recognized from the geological records, depending on the data type available for analysis. Domingo et al. (2014) inferred tectonic and climatic drivers over macroevolutionary patterns in mammals, corresponding to various geographic scales. Zuschin et al. (2014) concluded that understanding the ecological dynamics over geological time scales requires paleontological data in a sequence stratigraphic framework, which allows evaluation in an environmental context. The study had also stated that in aquatic habitats, water depth and hydrodynamic energy gradients influence dominant controls over species occurrence and diversity. These gradients show subtle long-term trends, corresponding to the sequence stratigraphic architecture. Tectonics affected the sequence architecture in this particular marginal marine setting: it controlled accommodation space and sedimentary input, and provided stable boundary conditions over hundreds of thousands of years. Application of sequence stratigraphic concepts, particularly the sequence development and sequence types to the documentation of habitat dynamics in geological history provides certain generalizations. The sequences formed under the influence of sea-level cycles show hierarchical stacking pattern, and this pattern was recognized by many authors in sedimentary facies successions (Goldhammer et al., 1991; McGhee, 1992; Gjelberg and Steel, 1995), geochemical profiles (Veizer, 1985; Veizer et al., 1997, 1999, 2000), lithofacies-based paleobathymetric trends (Ramkumar et al., 2004), and in multiproxy estimates (Ramkumar et al., 2011; Ramkumar, 2015). The sequence stratigraphic concepts (Vail et al., 1977) and attendant relative sea-level cycles (Haq et al., 1987, 1988) were introduced based on the stratal patterns as inferred from marginal siliciclastic deposits and were linked to the interplay between sediment supply-relative sea-level oscillations-climate-tectonics (Sarg, 1988; Grammer et al., 1996) which exhibit cyclicity.

The sequence stratigraphic concepts and the relative sea-level cycles (ranging in duration from few tens of millions to few million years, i.e., first to third order cycles) published by Vail et al. (1977) and revised by Haq et al. (1987, 1988) and updated more recently (in part) by Haq (2014) recognized about 100 global sea-level changes. Mitchum and Van Wagoner (1991) reported that the interpreted eustatic cyclicity has a pattern of superposed cycles with frequencies in the ranges of $9-10$, $1-2$, $0.1-0.2$, and $0.01-0.02$ Ma (second—fifth-order cyclicity, respectively). The studies, which followed since then, have also established that the earth processes were on operation at various timescales and generated unique hierarchical sedimentary facies succession principally under the influence of global sea-level cycles. Concise review on sequence stratigraphic methods and nomenclature can be found in Catuneanu et al. (2011), while Catuneanu (2006) presents a detailed account.

3.3 TYPES OF DEPOSITIONAL SYSTEMS

The responses of siliciclastic and carbonate systems to the interplays between prime controlling factors differ (Sarg, 1988), and there are differences in terms of habitat dynamics as well (Holland and Christie,

2013). While the siliciclastic system depends on sediment influx from land mass (Sandulli and Raspini, 2004), source rock composition, source area weathering, sorting, tectonic setting, diagenesis, and recycling (Alvarez and Roser, 2007), the carbonate system relies on the conditions of sediment production (Sarg, 1988). The environmental conditions within a carbonate-depositional system depend on the climatic, tectonic, and other extraneous factors too (Menier et al., 2014b). Further, the periods of significant carbonate accumulation accompany enhanced burial of organic matter predominantly originated from primary production within the basin and explain the enhanced marine biodiversity during periods of extensive carbonate deposition (Wilson, 2008). The mixed siliciclastic–carbonate system, the result of the interplay between the hydrodynamic factors affecting a particular environment (Menier et al., 2016) and the sediment-forming minerals (Garcia et al., 2004) is thus, far more complex than the siliciclastic and carbonate systems. Added complexity is the myriad of biotic response to specific conditions of environments (Traini et al., 2013; Montagne et al., 2013). For example, the larger benthic foraminifera shows tolerance to environmental conditions and occurs in shallow, tropical marine habitats and mixed carbonate–siliciclastic settings (Novak and Renema, 2015).

3.4 FACIES TYPES AS PROXIES FOR HABITAT HETEROGENEITY

As explained by sequence stratigraphic concepts and innumerable case studies that demonstrated the control exercised by sea-level fluctuations over spatiotemporal variations of facies distributions, it is perfectly logical to link the sea-level fluctuations resulting in a chain of responses in terms of changes in facies distribution, environmental parameters, and ecospace to finally affect the species occurrence, distribution, and population (Crampton et al., 2011; Sessa et al., 2012). If so, as demonstrated by the case studies cited, it becomes plain that, systematic documentation of facies distribution at spatiotemporal scale, with examination of sea-level trends would explain the magnitude of ecospace/habitat dynamics.

A link between habitat heterogeneity and taxonomic diversity is another way in which the geologic processes might directly influence living organisms to yield correlated patterns in the rock and fossil records. Zuschin et al. (2014) recorded perfect parallels between transition of depositional bathymetry and hydrodynamic energy that resulted in similar gradual transition in biotic response. McMullen et al. (2014) reported strong facies control on fossil concentrations, and their association with maximum flooding surfaces and lag deposits above sequence boundaries. Peters (2008) demonstrated the relationship between Phanerozoic marine biodiversity and the temporal and spatial distributions of carbonate and siliciclastic lithofacies. Smith et al. (2006) demonstrated the existences of marked differences in the composition and diversity of faunas across the shelf at a single-time interval and through time at the same locality, driven primarily by factors such as sedimentary facies, which are controlled by changing sea-levels. Sessa et al. (2012) interpreted that K/Pg extinction caused significant restructuring of the ecological composition of offshore assemblages and also that the ecological effects were facies specific, with offshore faunas displaying dramatic reorganization and shallow subtidal biota remaining relatively unchanged across the extinction. While examining the causes and events that took place during the Triassic-Jurassic transition, Ritterbush et al. (2014) reported that the observed facies changes previously attributed to sea-level change were interpreted to have resulted from the collapse of the carbonate factory concomitant with the mass extinction, with transition to an alternate state dominated by siliceous sponges before a return to carbonate platform development. This is one of the examples demonstrating the link between sea-level fluctuation-biotic response-habitat change and facies succession. It also supports the premise of documenting facies successions as proxy to track prevalent habitat diversity.

Smith and Benson (2013) studied the habitat diversity trends in relation to marine bedrock area, lithofacies diversity, and echinoid species diversity through an 80 Ma sea-level cycle, as recorded in the Cretaceous of the United Kingdom. They found positive correlation between bedrock area and number of lithofacies units with species diversity. They have also documented that there are no simple relationships between bedrock area and lithofacies diversity or the original area of marine sediments and range of habitats that once existed and also that these variables are strongly shaped by stratigraphic architecture. According to these authors, the geological record left by a major sea-level cycle accurately captures the biological evolution of local communities. It is incomplete and systematically biased in its coverage of environments and communities. Erosional truncation of sediment packages and species ranges in the later stages of sea-level cycles explains the reported asymmetric relationship. As the carbonate platforms are composed of facies mosaics that shift laterally in time and are frequently superimposed in the rock record, Darroch (2012) examined an important question of whether quantitative paleoecological studies focused on the fossil record in carbonate environments are subject to significant biases with successive vertical facies changes and found a strong taphonomic control on carbonate facies and faunal/habitat diversity. Brady (2016) was of the opinion that in addition to taphonomic control, the physical sedimentary environmental conditions such as sediment input, rates of physical and biogenic reworking, winnowing, and sediment production were also found to exercise control over diversity trends, which in turn vary according to habitat types, as examined from shallow to deep subtidal environments.

In summary, sequences are those facies bundles, bounded on either side by unconformities or correlative surfaces and created in response to relative sea-level fluctuations (Vail et al., 1977; Catuneanu, 2006; Catuneanu et al., 2011). The relative sea-level change, depending on the direction (shoreward or seaward) moves the principal loci of deposition and changes the types of facies bundles. The facies characteristics, thickness, and areal distribution of these bundles depend on a variety of factors including, tectonics, climate, duration and magnitude of sea-level fluctuation, sediment influx/production, initial topography of the depositional loci and its successive evolution, and others. Following the definition of sedimentary environment by Reineck and Singh (1980) which says that *sedimentary environment can be defined by its physical, chemical and biological conditions in a geomorphic setup*, we interpret the facies bundle in its least measured scale (dependent on the scale of interest; also explained and demonstrated in successive chapters) on spatial–temporal scale as unique habitat. Wilson (2015) recognized individual subenvironments as unique habitats and documented unique faunal diversity and population. This study stands testimony to the applicability and practicality of facies bundle as unique subenvironment with a definitive geomorphic attribute that can be recognized from geological records through proxies. In order to demonstrate these premises presented in this summary, idealistic models on sea-level change (fall and rise) and resultant shift of shoreline (seaward and shoreward) change in habitat distribution and shift of principal loci of deposition are presented in Fig. 4.4 and 4.5.

CHAPTER 4

The Cauvery Basin of South India: A Test Case

4.1 JUSTIFICATION AND OBJECTIVES

The Cauvery Basin (Fig. 4.1), one of the most studied basins of India (Acharyya and Lahiri, 1991), composes a variety of depositional systems *a la* micro–meso habitats as represented by alternations of siliciclastic–carbonate dominated cyclic facies bundles. Within the major siliciclastic–carbonate dominated depositional systems, a wide variety of lithological types deposited by high-frequency relative sea-level fluctuations occur. The fluctuations are in the order of 10^4–10^6 years and found to be consistent with the timescale-sea-level curve of Gradstein et al. (2004). All the six global sea-level peaks of Barremian–Danian are reported from this basin. The corresponding sequences are separated by type 1 sequence boundaries bounded within third-order sea-level cycles (Ramkumar et al., 2011). Probably resultant, significant enrichments and depletion of marine, coastal, and terrestrial biota in thick populations are also observable in the basin. These offer a test case of examining the influences of short- and long-term sea-level fluctuations on the habitat and biodiversity dynamics, and the abundance–diversity relationships. Thus, the objectives are set to (1) examine the relationship between the relative sea-level fluctuations and the habitat heterogeneity as evident from facies types, association and succession, (2) unravel the scale of changes in habitat (spatial and temporal), (3) document the associated biotic response in terms of occurrence, relative abundance, and diversity, (4) understand the potential controlling factors that may have influenced the habitat-biodiversity trends, as a result of climatic (warm and cool) conditions, sea-level lows and highs and different environmental settings (coastal, marginal marine, shelfal, and bathyal), and (5) examine the habitat-biodiversity trends of different chronological periods of the basin history (Barremian–Danian).

Eustasy, High-Frequency Sea-Level Cycles and Habitat Heterogeneity.
DOI: http://dx.doi.org/10.1016/B978-0-12-812720-9.00004-8

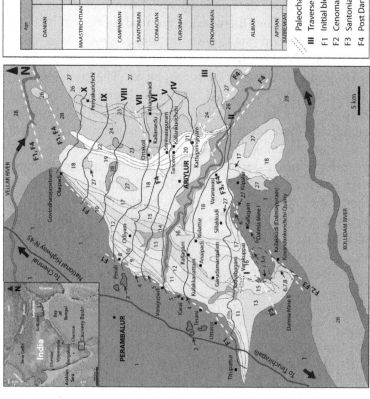

Figure 4.1 *Location of the Cauvery Basin and distribution of lithostratigraphic units.* The distribution pattern of lithostratigraphic members of the basin, their general younging nature toward east, contact relationships, and the tectonic and sedimentary structural information suggest that initial faulting during basin formation and reorganization during Santonian were the major tectonic events that impacted the depositional systems while relative sea-level fluctuation controlled the occurrence, distribution and survival of depositional systems all through the depositional history. After Ramkumar, M., Stüben, D. and Berner, Z., 2004a Lithostratigraphy, depositional history and sea-level changes of the Cauvery Basin, southern India. Annals of Geology of Balkan Peninsula 65, 1–27.

4.2 REGIONAL GEOLOGICAL SETTING

The Cauvery Basin (Fig. 4.1) is located between latitudes 08°30′N and longitudes 78°30′E in the south Indian Peninsula. It covers an exposed area of about 25,000 km^2 onland and 17,500 km^2 in offshore regions (Sastri et al., 1981) of the Bay of Bengal up to 200 m isobath. It is the southernmost basin among the NE–SW trending Late Jurassic-Early Cretaceous pericratonic rift basins (Sastri et al., 1981; Powell et al., 1988; Chari et al., 1995) formed all along the eastern continental margin of the Peninsular India (Ramkumar et al., 2016a). This basin was created during Late Jurassic-Early Cretaceous (Powell et al., 1988; Chari et al., 1995; Jafer, 1996; Chatterjee et al., 2013) fragmentation of Gondwana super continent and rifting between Africa–India–Antarctica (Lal et al., 2009; Ramkumar et al., 2016a) and continued evolving till the end of Tertiary (Prabakar and Zutshi, 1993; Ramkumar et al., 2016a, b). The sedimentary succession of this basin exceeds 5500 m in thickness (Govindan et al., 2000). Comprehensive lithostratigraphy of the onland part this basin was presented by Ramkumar et al. (2004a). The lithostratigraphic subdivisions (Table 4.1; Figs. 4.1 and 4.2) are separated by sequence boundaries and other correlative surfaces (Ramkumar et al., 2011) and are geochemically distinct to the tune of 100% from each other (Ramkumar et al., 2010b). The evolutionary (Sastri et al., 1981; Prabhakar and Zutchi, 1993; Chari et al., 1995; Lal et al., 2009; Ramkumar, 2015; Ramkumar et al., 2016a,b), stratigraphic (Ramanathan, 1968; Banerji, 1972; Sundaram et al., 2001; Ramkumar et al., 2004a, 2005a), paleontologic (Chiplonkar, 1987; Govindhan et al., 1996; Bhatia, 1984; Jafer, 1996; Jafar and Rai, 1989; Kale and Phansalkar, 1992a, b; Kale et al., 2000; Guha, 1987; Guha and Senthilnathan, 1990, 1996; Ramkumar and Chandrasekaran, 1996; Ramkumar and Sathish, 2009; Ramkumar et al., 2010a; Rai et al., 2013; Paranjape et al., 2014), and geochemical (Ramkumar et al., 2004b, 2005b, 2006, 2010b, 2010c, 2011) traits of this basin are excellently documented.

4.3 MATERIAL AND METHODS

Systematic field mapping at the scale of 1:50,000 was conducted through 10 traverses (Fig. 4.1). A total of 308 locations were logged and sampled. Information on lithofacies, contact relationships,

Table 4.1 Lithostratigraphy of the Exposed Part of the Cauvery Basin

Age	Formation	Member	Thickness (m)
Mio- Pliocene	Cuddalore S.St. Fm.		>150
	------------------------------*Unconformity*------------------------------		
	Niniyur Fm.	Periyakurichchi biostromal Mbr.	26
Danian		Anandavadi arenaceous Mbr.	30
	------------------------------*Unconformity*------------------------------		
	Kallamedu Fm.		100
	------------------------------*Unconformity*------------------------------		
	Ottakoil Fm.		40
Maastrichtian	------------------------------*Unconformity*------------------------------		
		Srinivasapuram gryphean L.St. Mbr.	18
	Kallankurichchi Fm.	Tancem biostromal Mbr.	8
		Kattupiringiyam inoceramus L.St. Mbr.	8
		Kallar arenaceous Mbr.	6
	------------------------------*Unconformity*------------------------------		
Campanian		Varanavasi S.St. Mbr.	270
	Sillakkudi Fm.	Sadurbagam pebbly S.St. Mbr.	80
Santonian		Varakuppai lithoclastic conglomerate Mbr.	45
	------------------------------*Unconformity*------------------------------		
Coniacian		Anaipadi S.St. Mbr.	215
	Garudamangalam Fm	Grey S.St. Mbr.	35
		Kulakkanattam S.St. Mbr.	123
	------------------------------*Unconformity*------------------------------		
Turonian	Karai Fm.	Odiyam Sandy clay Mbr.	175
		Gypsiferous clay Mbr.	275
	------------------------------*Unconformity*------------------------------		
Cenomanian		Kallakkudi Calcareous S.St. Mbr.	60
	Dalmiapuram Fm.	Olaipadi conglomerate Mbr.	65
		Dalmiya biohermal L.St. Mbr.	15
Albian		Varagupadi biostromal L.St. Mbr.	23
		Grey shale Mbr.	7
	------------------------------*Unconformity*------------------------------		
Aptian		Terani clay Mbr.	30
	Sivaganga Fm.	Kovandankurichchi S.St. Mbr.	24
Barremian		Basal Conglomerate Mbr.	18
	------------------------------*Unconformity*------------------------------		

Basement Rocks (Granitic gneiss, charnockite, pegmatite, etc.)

The basin-fill in the Cauvery Basin accounts for about 5500-m-thick stratigraphic records of Barremian–Danian. Stratigraphic records in onland counterpart/extension expose all of them with reduced thicknesses.

After Ramkumar, M., Stüben, D. and Berner, Z., 2004a Lithostratigraphy, depositional history and sea-level changes of the Cauvery Basin, southern India. Annals of Geology of Balkan Peninsula 65, 1–27.

sedimentary, and tectonic structures and occurrences of mega and ichnofossil assemblages, their relative diversity, and population were recorded at each location logged. Based on the information, a composite stratigraphic profile of Barremian–Danian strata was constructed that allowed facies and petrographic analyses. It was followed by interpretation of environmental conditions of each

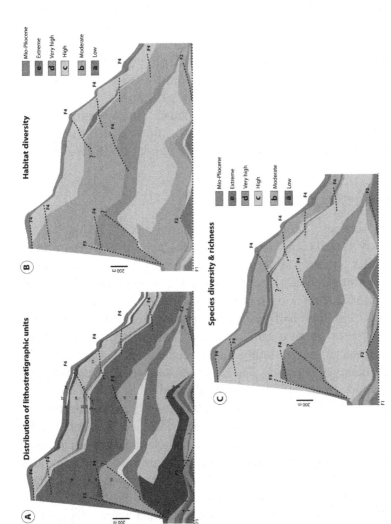

Figure 4.2 **Spatiotemporal variations of lithostratigraphic units along the traverses.** (A) depicts the lithostratigraphic members as mapped along the traverse lines presented in Fig. 4.1. It also provides a generalized scale of shift in major habitats during Barremian—Danian in the Cauvery Basin, as a function of relative sea-level fluctuations than tectonic events. Owing to the configuration of the basin and confined nature of provenance, other factors namely, climate and sediment influx were subdued and or acted in resonance with the eustasy and basin-scale tectonics. Depictions of changes in species diversity, population, and habitat dynamics in terms of a relative gradational scale are also provided in this figure. Relative changes of the habitats through time are summarized in (B) and the species diversity and richness are summarized in (C). Comparison of these two figures suggests a disparity between paleobiodiversity and habitat heterogeneity trends. It also paves way for the interpretation of involvement of multiple environmental and other parameters that influenced habitat and biodiversity trends.

facies type recognized, and documentation of stratigraphic variation of environmental settings. Compilation of these inferences together with ecological/environmental preferences of fauna/flora recorded in the succession and the lifestyle of the biota allowed interpretation of stratigraphic variations of relative bathymetry, habitat heterogeneity, and diversity. In addition to the field data on the occurrences of mega and ichnofossil assemblages, their relative abundances, data from published reports, papers, and open source fossil databases were also consulted to compile information on chronological variations of fossil type, population, diversity, and others. The information were then analyzed in the context of known and inferred geological events that took place in the basin to constrain on the relative stability/heterogeneity of habitats under the influence of sea-level fluctuations and resultant species abundance and diversity.

4.4 TECTONIC STRUCTURES

The rift between India—Australia—Antarctica during Late Jurassic—Early Cretaceous resulted in block faulting of Precambrian terrain of India, creating a series of sedimentary basins (Ramkumar et al., 2016a) along the east coast, among which the Cauvery Basin is the southernmost sedimentary basin. It is a structurally elongated basin with NE—SW trending half-graben morphology and a regional dip averaging 5—10° to the east and southeast. Toward west, it is demarcated by steep, basin margin faults, which separate the Archaean shield from the sedimentary deposits. The basement tectonics, as enumerated by Prabakar and Zutshi (1993), and Watkinson et al. (2007) and later detailed by Ramkumar et al. (2004a, 2005) and Ramkumar (2015), stated that since initial rifting, this basin continued to evolve until the end of Tertiary through rift, pull-apart, shelf sag, and tilt phases, accompanied by episodes of transgression, regression, erosion, and deposition to fill the basin (Ramkumar et al., 2004a). Seismic sounding and exploratory borehole log data revealed that this basin consists of a number of subbasins, namely, Ariyalur—Pondicherry depression, Tanjore—Tranquebar—Nagapattinam depression, and Ramnad—Palk Bay depression that are differentiated by many highs, namely Kumbakonam—Mandanam ridge, Pattukottai—Mannargudi—Karaikal ridge, and Mandapam—Delft ridge (Sastry et al., 1977; Kumar, 1983; Acharya and Lahiri, 1991; Chandra, 1991; Prabhakar and Zutshi,

1993; Chari et al., 1995). Among these, the Pondicherry subbasin is the northernmost and contains three main exposures mappable in outcrops, in which the region confined between Vellar River in the north and Coleroon River in the south is the largest and exposes Barremian–Danian strata. The tectonic structural features mapped in the field are presented in Fig. 4.1. Based on field criteria, lithological association and displacement, the basin-scale tectonic movements were interpreted, namely, initial block faulting (as indicated by the northeast to southwest fault lines in the basement rocks all along the western margin of the basin) during Barremian (Refer F1 in Fig. 4.1), movement of fault blocks during Cenomanian (F2 in Fig. 4.1), reactivation of older fault blocks and creation of new fault during Middle Santonian (F3 in Fig. 4.1), and reactivation of fault blocks during post-Danian–pre-Quaternary (F4 in Fig. 4.1). In addition, there were minor and local-scale tectonic movements, all of which confined only to the adjustment of fault blocks along the preexisted fault planes. The post-Danian faults that affected the Miocene–Pliocene sandstones, Danian limestones, and Maastrichtian deposits show differences in trends from that of pre-Danian faults, as depicted in Fig. 4.1.

4.5 LITHOFACIES SUCCESSION AND DISTRIBUTION

A brief description on facies characteristics of the Barremian–Danian lithofacies succession is presented in Table 4.2. Fig. 4.2 depicts the distribution, contact relationships, and thicknesses of lithostratigraphic formations and members as observed along the 10 traverses (refer Fig. 4.1). The facies succession of this basin, as could be appreciated from Table 4.2 and Fig. 4.2, are essentially alternations of siliciclastics and carbonates. Yet, the siliciclastic deposition dominated the depositional history (temporal) as well as areal extent (spatial). It is also explicit from Table 4.2 and Fig. 4.2 that the initiation of siliciclastic deposition was either commenced due to ravinement events (as a result of eustatic sea-level lowering) and or major tectonic events (tectonoeustasy—refer Fig. 4.3), while the carbonates were associated with conformable contacts with underlying deposits, maximum flooding surfaces and inherited preexisted depositional topography, implying essentially depositional events following eustatic sea-level rise. Superimposed over this long-term trend, occurrences of contacts between very old deposits and much younger deposits in the lower half of the succession and absence of such

Table 4.2 Lithofacies Characteristics of the Barremian–Danian Strata of the Cauvery Basin

Formation	Member	Facies Characteristics
Niniyur	Periyakurichchi	Thick arenaceous bioclastic limestone bed followed by medium to thick, parallel, even-bedded, recurrent biostromal limestone and marl typify this member. Regionally varying concentrations of shell fragments and whole shells of bivalve, gastropod, and remains of amphibia, pisces, algae, foraminifera, and ostracoda are also observed
	Anandavadi	Isolated coral mounds, impure arenaceous limestone, and lenses of sandstone and clay, deposited in a restricted marine, subtidal to intertidal regions. Occurrence of localized concentrations of shell fragments, coralline limestone and reef derived talus deposits are characteristics of this member. This member rests over the Kallamedu Formation with distinct disconformity. At top, an erosional surface is recognizable
Kallamedu		Unconsolidated, well-rounded, and poorly-sorted barren sands with rare-scarce dinosaurian bone fragments. Toward top, these grade to medium to thin bedded, relatively highly argillaceous sandstones. Local occurrence of clays and silts with dispersed detrital quartz grains and sandy streaks, nonbedded nature, rare lamination, and mud cracks in them indicate sedimentation as over bank deposits. Toward top, this formation has paleosol indicating return of continental conditions
Ottakoil		The rocks are coarse to medium sized, well-sorted, fossiliferous, low-angle cross-bedded, and planar to massive bedded sandstones with regionally varying sparse calcareous cement. They also show recurrent fining upward sequences. Abundant *Stigmatophygus elatus* and few trace fossils indicate shallow marine environment of deposition. This formation rests over the Kallankurichchi Formation with disconformity and overlapped by the Kallamedu Formation
Kallankurichchi	Srinivasapuram	Uniform, parallel, thick-very thick-bedded gryphean shell banks with *Terebratula*, *Exogyra*, bryozoa, and sponge. Extensive boring in gryphean shells, synsedimentary cementation, colonies of encrusting bryozoa over gryphean shells, and micritization of bioclasts indicate deposition in inner shelf
	Tancem	Biostromal limestone beds with thin to thick, parallel, even-bedded nature, cross-bedding, normal grading, hummocky cross-stratification, feeding traces, escape structures, and tidal channel structures. Local concentrations of various fossils, sporadic admixture of siliciclastics, and intraclasts are frequently observed.
	Kattupiringiyam	Dusty brown, friable, carbonate sands with parallel, even, and thick to very thick bedding that contain only *Inoceramus* and bryozoa. This member has diagenetic bedding and abundant geopetal structures filled with mm to cm sized dog tooth spars of low magnesian nonferroan calcite. This member has nondepositional surface at bottom and has erosional surface at top
	Kallar	Normal-graded conglomerates in which well-rounded clasts of basement rocks, fresh feldspar, resedimented colonies of serpulids, and other older sedimentary rocks that range in size from coarse sand to boulder are observed. Lower contact of this member is an erosional surface. The upper contact is nondepositional surface resulted from marine flooding

(Continued)

Table 4.2 (Continued)

Formation	Member	Facies Characteristics
Sillakkudi	Varanavasi	Featureless, massive, thick to very thick bedded, coarse to medium grained sandstones. Occasional very coarse sandstone lenses, pockets of shell hash, intraformational lithoclastic bounders in association with vertical cylindrical burrows, and resedimented petrified wood logs are observed. The rocks rest over the pebbly sandstone member with nondepositional surface. Upper surface of this member represents erosional surface associated with regression
	Sadurbagam	Coarse siliciclastics with abundant marine fauna, shell fragments and varying proportions of calcareous matrix. At the base, an erosional surface followed by distinct cobble–pebble quartzite conglomerate is observed. The rocks show normal grading, low angle cross-bedding, massive, thick to medium, even and parallel bedding. At places, pockets of shell rich carbonate lenses with abundant siliciclastic admixture are found to occur. Load casts, slump folds, pillow structures, and synerasis cracks, occasional development of algal mounds are also found
	Varakuppai	It rests over older sedimentary rocks with typical erosional surface. The erosional intensity was such high that, the beds have direct contact with much older Karai Formation. Fluviatile sandstones with well-rounded basement rocks, quartzite, and older sedimentary rock boulders in addition to unsorted coarse sand–pebble-sized siliciclastics constitute this member. These are typically reverse-graded and show cyclic bedding, large-scale cross bedding and lack any body fossils. Large-scale cross bedding, mud drapes, fresh feldspar, and sandstone clasts are also recorded. Toward top, *thalassinoid* burrows are reported, that indicate gradual submergence of the depocenter by rising sea-level
Garuda mangalam	Anaipadi	Massive and thin-bedded claystones, silty claystones, and clayey sandstones in south that gradually grade to silty clay in south center and thin down. Again, from there, thickness of these beds and sediment grain size increase and contain abundant large ammonites. Further north, these were observed to be clayey siltstones with abundant shell fragments and ammonites
	Grey sandstone	Highly well-cemented, sorted and rounded grains giving massive appearance. The beds are cyclic, parallel, even-bedded alternative layers of barren and highly fossiliferous and sandy layers with regionally varying thicknesses. This member rests conformably over the Kulakkanattam Member and has distinct erosional and nondepositional surface. Upper contact is nondepositional surface associated with marine flooding
	Kulakkanattam	Massive, yellowish brown, ferruginous sandstones with abundant admixture of silt and clay. Localized concentrations of shell fragments, bivalves and gastropods, and ammonites are common. It also contains abundant wood fragments with extensive oyster boring. Cross-bedding, channel courses, planar-bedding and feeding traces are also common. The depositional surface was strongly bioturbated and riddled by roots. An angular erosional unconformity separates this member from underlying Karai Formation
Karai	Odiyam sandy clay	Silty clays and sandy clays with abundant ammonites. Load structures and syndepositional slump folds are frequently observed. Upper portion of this member has localized pockets of fine sandstone along with ammonites. While the lower contact is conformable with underlying member, upper contact is erosional

(Continued)

Table 4.2 (Continued)

Formation	Member	Facies Characteristics
	Gypsiferous clay	Unconsolidated deep marine clays and silty clays. These beds contain thick population of belemnite rostrum and phosphate nodules. While a nondepositional unconformity surface separates this member from the underlying member, upper contact is nondepositional, and erosional. From south to north, gradual reduction of thickness, population of belemnite, and phosphate nodules and frequency of gypsum layers are observed. Repetitive occurrences of 1–3 cm thick red and green colored clay layers that can be traced for many kilometers are ubiquitous
Dalmiapuram	Kallakkudi	Fine-coarse sandstones with alternate medium to thick beds of silty clay, calcareous siltstones, bioclastic arenaceous limestone, and gypsiferous clay. In the southern region, these beds show recurrent bands of fining upward sequences of siliciclastics with calcareous cements. The intercalations are recurrent and show typical Bouma sequences, normal-grading, load casts, and channel and scour structures. Toward northern regions, this member grades to more silty and clayey, but gradation and gypsiferous bands are persistent with an addition of ferruginous silty clay bands
	Olaipadi	Basinal silty clays and clays in which chaotic blocks are embedded. The beds contain large blocks of angular and subrounded basement rocks, coralline limestone, claystones (lithoclasts of carbonates and greenish claystones typical of Karai Fm.) and lithoclasts of older conglomerates, and others. Toward the top, deep marine clays grade into calcareous siltstone and contain granitic cobbles and minor amounts of siliciclastic sands
	Dalmia	Pure algal and coral facies limestone beds that form reef core. Upper contact of this member is a forced regression surface
	Varagupadi	Limestone beds typical of reef flank biostromal beds deposited under high-energy conditions. Thin to thick bedded, even to parallel, bioclastic limestone beds that have drawn their detritus from reefs predominate. These beds are found to be directly overlying the Grey Shale Member. The rocks show wackestone to rudstone fabric and have clasts of redeposited boundstones
	Grey shale	Grey shale beds with frequent thickening upward interbeds of fossiliferous grey limestone and minor to significant admixture of silt sized siliciclastics. Lower contact of this member is an unconformity surface associated with marine flooding and the upper contact is nondepositional and erosional
Sivaganga	Terani Clay	White to brownish colored clay and argillaceous siltstone that show transition from Kovandankurichchi Member. Beds are massive to very thick in nature. Lower contact of this member is nondepositional surface
	Kovandankurichchi	Grain supported coarsening upward cyclic beds (20–100 cm thick each) of very coarse sandstones that show parallel, even and thin to thick bedding. Grains are well sorted within each lamina and show rounded-well rounded shape. These represent recurrent sheet flow deposits probably in a subaqueous fan deltaic environment
	Basal conglomerate	Recurrent fining upward sequences of lithoclastic conglomerates of fluviatile and coastal marine environments. Lithoclasts are of gneissic basement rocks. Rests over basement rocks with distinct erosional surface. Upper contact is a nondepositional surface

From the facies characteristics, in addition to the cyclic occurrences of carbonates and siliciclastics, it can also be inferred that the Cauvery Basin was starving for sediments since inception. The carbonate deposition commenced whenever siliciclastic influx dwindled or ceased to reach depositional sites. A progressive stability of depositional environments can also be perceived from older to younger strata.

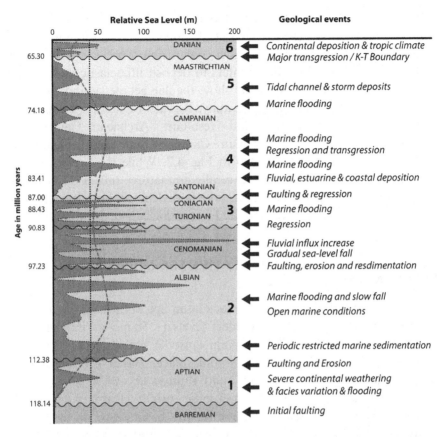

Figure 4.3 **Chronological depiction of Relative sea-level fluctuations and major tectonoeustatic and depositional/** **erosional events during Barremian–Danian in the Cauvery Basin.** *Solid line curve is indicative of absolute values of Relative Sea-Level (RSL), while the dotted line (. . .) indicates linear trend and the dashed line (----) indicates polynomial trend. Recognizable sea-level cycles are indicated by Arabic numerals. The sea-level curve was constructed based on bathymetric estimates of facies types and supplemented with biofacies. Independent paleobathymetric estimates based on foraminifer also affirmed the trends. Further, the curve documents all the six eustatic sea-level peaks. Major shoreline shifts and tectonic events are also indicated in the figure, from which, resultant temporal changes in habitats can be perceived.* After Ramkumar, M., Stüben, D. and Berner, Z., 2004a Lithostratigraphy, depositional history and sea-level changes of the Cauvery Basin, southern India. Annals of Geology of Balkan Peninsula 65, 1–27.

features in younger strata indicate subdued influence of tectonics during latter part of the evolutionary history of the basin. It is also discernible from Fig. 4.2 that the depositional topography and the lithofacies architecture continually varied until the end of Santonian. It was due to the fact that a major tectonic movement during Santonian that caused transgression in areas that remained topographically highlands since inception of the basin (Barremian) changed the depositional pattern from dynamic depocenters to more stable shelf conditions (Ramkumar et al., 2005a). At each major unconformity, basement rocks and older sedimentary rocks were eroded and transported to newer depocenters to deposit lithoclastic conglomerates. The post-Santonian deposits show regular geometry and followed the preexistent depositional topography. Prevalent extensive ravinement and shelf erosion events and resultant Archaean–Campanian, Turonian–Santonian, and Turonian–Campanian are also explicit from the contact relationships depicted in Fig. 4.2. While the pre-Santonian deposits are characterized by texturally immature, mixed siliciclastics and argillites, the younger (post-Santonian) deposits are characterized by carbonates, textural inversion, and mineralogical–textural maturity.

4.6 BARREMIAN–DANIAN RELATIVE SEA-LEVEL FLUCTUATIONS

Sedimentation in this basin took place in an epicontinental sea and the bathymetry was at shallow—modest levels (<50 m—as indicated by the linear curve in Fig. 4.3) wherein episodic bathymetric variations from supratidal or basinal levels took place. Based on foraminifer data, Raju and Ravindran (1990) and Raju et al. (1993) documented six third-order cycles of glacio-eustatic origin. Ramkumar et al. (2004a) constructed a sea-level curve for this basin based on bathymetric determinations of lithofacies, microfacies, and facies zone characteristics, synsedimentary structures, and mega and ichnofaunal assemblage data. It has recorded fourth and higher-order sea-level cycles (Fig. 4.3). The global sea-level peaks during 104 Ma (Early to Late Albian), 93.7 Ma (±0.9; Middle to Late Cenomanian), 92.5 Ma (±1; Early to Middle Turonian), 86.9 Ma (±0.5; Early to Late Coniacian), 85.5 Ma (±1; Early to Late Santonian), 73 Ma (±1; Late Campanian), 69.4 Ma (Early to Late Maastrichtian), and 63 Ma (±0.5; Early to Middle Danian) occurred in this basin (Raju and Ravindran, 1990; Raju et al., 1993; Ramkumar et al., 2004a). The third-order cycles are separated by type I sequence boundaries (recognized through shift of shoreline crossing shelf break as explicit in lithologic information, contact relationship between strata, evidences of subaerial exposure, and erosion, advancement of fluvial channels over former offshore regions, etc.). On a basin scale, the duration from

Barremian to Coniacian experienced high-frequency/higher-order cycles while Coniacian to Danian experienced sea-level rise and fall punctuated with lesser frequency of higher-order cycles. Steadily increase of sea-level is recognizable during Santonian–Early Campanian. In this backdrop, a gradual reduction of sea-level (Fig. 4.3) from Albian to Santonian and then a gradual rise up to the end of Cretaceous followed by a sea-level fall until Middle Danian and finally rise of sea-level could be observed. These trends are consistent with the sea-level curve of Gradstein et al. (2004) and recent revision of Cretaceous sea-level chart by Haq (2014). Corroboration of this curve with facies characteristics, contact relationships, tectonic, and sedimentary structures, fossil data, and others revealed that the sea-level fluctuations were primarily eustatic in nature and were affected by sporadic attenuation and or diminished by tectonoeustasy. These genetic characteristics of sea-level fluctuations and tectonoeustatic events are summarized in Fig. 4.3. Idealistic diagrammatic visualizations of these geological events (during sea-level rise—Fig. 4.4; during sea-level fall—Fig. 4.5) and selected stratigraphic successions (Cenomanian, Santonian–Campanian, Maastrichtian–Danian) are presented in Figs. 4.6–4.8.

4.7 BARREMIAN–DANIAN BIOTIC HETEROGENEITY

Biotic occurrence with reference to individual lithostratigraphic members is presented in Table 4.3. This list is indicative only and not a complete and exhaustive inventory. The list is made from the data collected during the field survey, laboratory examination, and from previous publications. Based on this list and the perceptions of previous publications enumerating biodiversity characteristics and evaluations (by authors of this book), relative population density and biodiversity were defined. Selected fossils, as recognizable in the field, hand specimen, and reflected and transmitted light microscopy, are presented in Plate 4.1. These information were combined with the sea-level fluctuations (Fig. 4.3), lithofacies variations (Figs. 4.2, 4.6–4.8), interpretations of habitat heterogeneity as inferred from field data, prevalent geological events, and visualizations on habitat change (shift, enlargement, reduction, destruction, etc.) were collated to interpret habitat dynamics and are listed in Table 4.3 and depicted in Fig. 4.2. The table shows that while the carbonate lithological

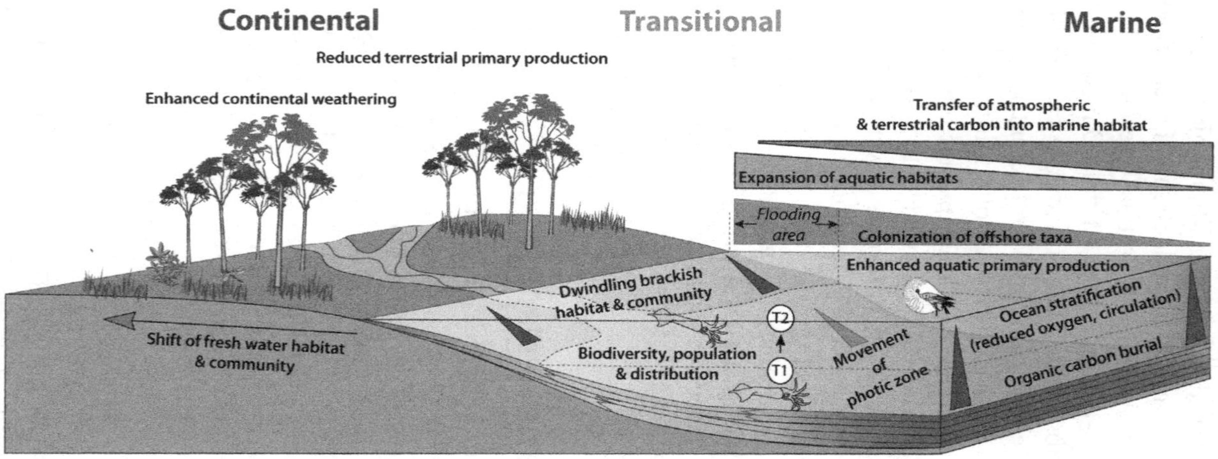

Continental

Reduced terrestrial primary production

Enhanced continental weathering

Shift of fresh water habitat & community

Transitional

Dwindling brackish habitat & community

Biodiversity, population & distribution

Flooding area

T1 T2

Marine

Transfer of atmospheric & terrestrial carbon into marine habitat

Expansion of aquatic habitats

Colonization of offshore taxa

Enhanced aquatic primary production

Movement of photic zone

Ocean stratification (reduced oxygen, circulation)

Organic carbon burial

*Figure 4.4 **Idealistic model of depositional system during sea-level rise.** Diagram represents sea-level rise (from T1 to T2) and associated changes in the oceanographic phenomena, influencing facies distribution and ultimately affecting the type, areal extent, distribution, and heterogeneity of habitat and biota.*

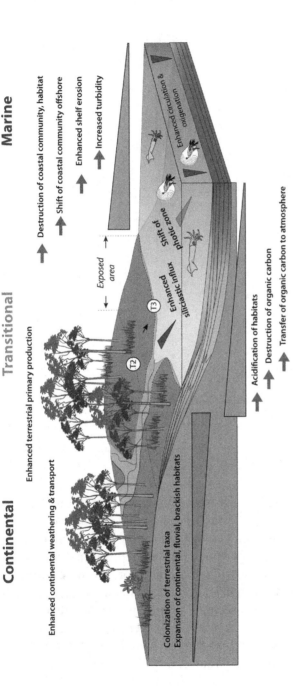

Figure 4.5 *Idealistic model of depositional system during sea-level fall (from T2 to T3) and associated changes in the oceanographic phenomena, influencing facies distribution and ultimately affecting the type, areal extent, distribution, and heterogeneity of habitat and biota.*

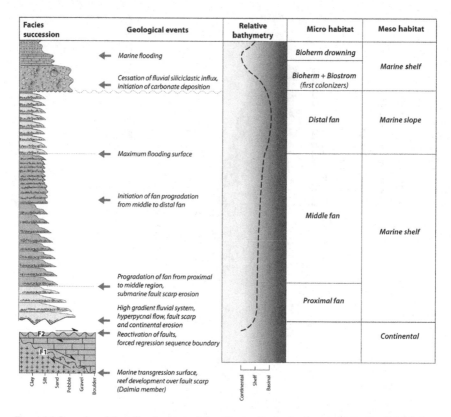

Figure 4.6 **Dynamics of depositional systems during Cenomanian in the Cauvery Basin.** *Detailed lithofacies succession is presented in the litholog. Depositional events in relation to these facies characteristics are also depicted at appropriate stratigraphic levels. From these, it is perceptible that, the siliciclastic domination, and coeval physical destruction due to syntectonics, the depositional system was not supportive of abundance of biota. The figure also depicts interpretation of meso and microhabitats from the facies characteristics and paleoenvironmental inferences.*

types show significant biodiversity and abundant population, the siliciclastics show comparatively lesser diversity and abundance. On a geologic time scale, consistency of habitat was not conducive of diversity and abundance. On a basin scale, the depositional as well as climatic conditions were highly fluctuating prior to the Santonian which are aptly reflected in higher habitat heterogeneity and species turnover and persistent and secular trends were prevalent since Santonian reflected in lower habitat heterogeneity, higher species diversity, and population. In this general backdrop, the very high fluctuations observed across the Cretaceous—Paleogene boundary are anomalous.

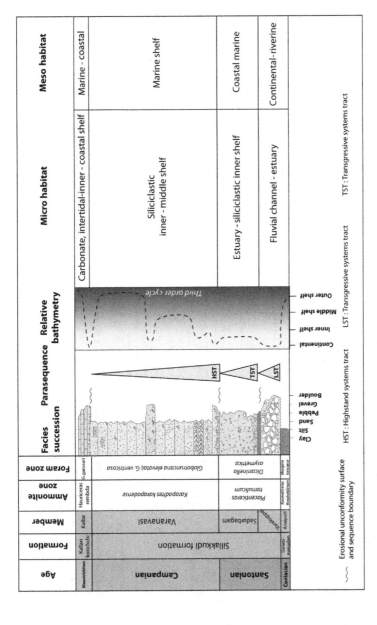

Figure 4.7 **Dynamics of depositional systems during Santonian–Campanian in the Cauvery Basin.** Detailed lithofacies is presented in the litholog. Depositional events in relation to these facies characteristics are also depicted at appropriate stratigraphic levels. The model depicts the tectonic reorganization of the basin, followed by invigorated fluvial channel that supported insignificant biotic accumulation, followed by gradual subsidence/sea-level rise leading to establishment of estuarine and coastal marine conditions that supported considerable biota and finally to the establishment of widest shelf in which abundant biota thrived. The figure also depicts compilation of micro and meso habitats from facies characteristics.

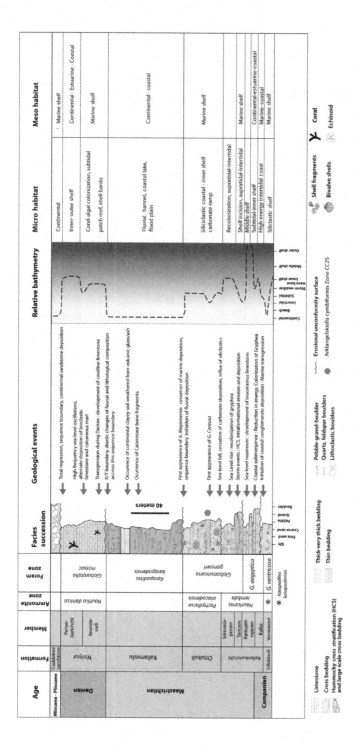

Figure 4.8 Dynamics of depositional systems during Late Campanian–Danian in the Cauvery Basin. The detailed lithofacies is presented in the litholog. Depositional events in relation to these facies characteristics are depicted at appropriate stratigraphic levels. The model depicts the tectonic quiescence and predomination of eustatic sea-level fluctuation and resultant facies/habitat change. The inferred micro and meso habitats are also depicted in the figure.

Table 4.3 Biodiversity and Habitat Dynamics during Barremian–Danian in the Cauvery Basin

Age	Fm.	Mbr.	Faunal and Floral Composition	Population Density	Diversity	Ecospace Dynamics
Danian	Niniyur	Periyakurinchi biostromal	Shark, algae, coral, bivalve, amphibia, pisces, gastropod, and *Hercoglossa danica*	Very high	Highly rich	Short-term cycles of sea-level fluctuations marked by marl-limestone alternations, preserving significant macro and microfaunal diversity and populations. While marl intervals document lesser diversity and population, the limestone intervals show abundance of population and species diversity
		Anandavadi arenaceous	Coral, algae, serpulid, bivalve, gastropod, and *Hercoglossa danica*	Very high	Highly rich	Typical coral-algal reef environment due to offshore protected environment. Isolated patch reefs represented by diverse coral colonies and associated algal and foraminiferal populations. Isolated bivalve shell banks also present with extremely rich populations
Maastrichtian		Kallamedu	Abelisauridae, Troodontidae, Sudamericidae, *Bruhatkayosaurus matleyi, Megalosaurus* sp. *cf. Simosuchus* sp., *Crocodyla, Kurnademys kallamedensis, Anura, Lepisosteidae, Aquilapollenites bengalensis, Cranwellia cauveriensis, Araucariacites australis, Tricolpites microreticulatus, Triporopollenites minimus, Discorhabdus ignotus, Eprolithus floralis, Pseudomicula quadrata, Holococcolith* sp., *Prediscosphaera* sp., *Uniplanarius* sp., *Petrobrasiella? bownii, Eprolithus* sp., *Corollithion exiguum, Ottavianus terrazetus, Lucianorhabdus cf. L. maleformis, Calculites obscurus, Ceratolithoides pricei,* and *Biantholithus cf. B. sparsus*	Nil to abundant	Nil to insignificant	A mix of estuarine and lagoonal nannofossil taxa and absence of marine communities

(Continued)

Table 4.3 (Continued)

Age	Fm.	Mbr.	Faunal and Floral Composition	Population Density	Diversity	Ecospace Dynamics
	Kallankurichchi	Ottakoil	*Pachydiscus otacodensis, Stigmatophygus elatus, Nautilus, Gryphaea, Alectryonia, Thalassinoides, Ophiomorpha, Dactyloidites Durania mutabilis, G. gansseri, Arkhangel-skiella cymbiformis, Braarudosphaerabigelowii, Ceratolithusaculeus, Chiastozyguslitterarius, Cyclagelosphaera deflandrei, Cribrosphaerella ehrengergii, Cribrosphaera sp.. Eiffelithusgorkae, E. parallelus, E. turriseiffeli, Microrhabdulusundosus, Macula decussata, M. staurophora, M. swastika, Petrobrasiella? bownii, Prediscosphaeracretacea, P. spinosa, Stradneriacrenulata, Staurolithitescrux, Zygodiscus minimus and Z. spiralis, and Agerostrea ungulate*	**Scarce to abundant**	**Rare to significant**	**Dwindling of faunal diversity and population in a waning sea. Coastal communities of macro and ichnofauna amid rich nannofossil assemblage**
		Sirivasapuram gryphean L.St.	*Gavelinopsis bembix, Gyroidinoides globosa, Lingulogavelinella, Cibicides, Siderolites, Pecten, Onychocellida, Hauriceras rembda, Perustrombus indicus, Gryphaea, Alectryonia, Terebratulida,* bryozoa, ostracoda, and sponge	**Extremely dense**	**Extremely rich**	**Extremely favorable environmental parameters such as warm, clear, turbid free, and circulated waters that were sustained for a long duration by stable sea-level. These provided for diverse and rich fauna and flora composition. Toward top, reduction of sea-level accompanied by increase of siliciclastic influx is noted, together with dwindling of fossil diversity and population**
		Tancem biostromal	*Stigmatophygus, Planolites, Ophiomorpha irregulaire,* and *Megalolithus cylindricus*	**High**	**Significant**	A significant regression and increase in episodic high energy conditions. Opportunistic colonizers, coastal communities of body and ichno taxa thrived

Stage	Formation	Member	Fauna			Remarks
Campanian	Sillakkudi	Kattupiringiyam inoceramus L.St.	*Gavelinopsis bembix, Gyroidinoides globosa, Lingulogavelinella, Agerostrea ungulata, Ceratostreon pliciferum Planospirites ostracina, Pycnodonte (Phygraea) vesicularis, Rastellum (Arcostrea) pectinatum, and bryozoa*	Extremely dense	Very rich	**Flooding of the depocenter and prevalence of stable, relatively deeper middle shelf conditions allowed predomination of inoceramus population, followed by formation of *Exogyra*, *Alectryonia* dominant shell banks and associated diverse macro and microfauna**
		Kallar arenaceous	*Gavelinopsis bembix, Gyroidinoides globosa, Lingulogavelinella, Cibicides, Siderolites, Gryphaea, Alectronia, and Pecten Eubaculites Orbitoides* spp. and *Goupillaudina daguini*	Very high	Very rich	**A major transgression covering partly the former shelf. Prevalence of turbulent energy conditions accompanied by very high rate of lithoclastic influx into the depocenter thwarted hospitable conditions during initial duration, which in turn was gradually transformed into hospitable during later part and allowed colonization by *Gryphaea* and *Alectryonia***
		Varanavasi S.St.	*Inoceramid, serpulid, Turritella, Ophiomorpha, Thalassinoides, Karapadites karapadense, Globotruncana arca, Globigerinelloides, Marginotruncana marginata, Whiteinella baltica, Archaeoglobigerina, Bolivinoides culverensis, B. decorates, Globotruncana elevata, and G. ventricosa, Ostrea zitteliana*	Scarce to abundant	Rich	**Stable shelf, episodically inundated by high-energy events. Warm, stable, normal marine conditions on a widest shelf conditions promoted thriving of organisms and their preservation**

(Continued)

Table 4.3 (Continued)

Age	Fm.	Mbr.	Faunal and Floral Composition	Population Density	Diversity	Ecospace Dynamics
		Sadurbagam pebbly S.St.	Algae, *Rhynchonella*, Terebratulid, *Nautilus, Inoceramus balitus*, echinoid, crinoid, and bivalve	Scarce to abundant	Rich	Fewer coastal communities and others often transported from offshore and accumulated. High energy siliciclastic environments restricted the diversity and population
Santonian	Garudamangalam	Varakuppai lithoclastic conglomerate	*Thalassinoides, Marginotruncana Coronate,* and *Dicarinella asymmetrica*	Nil to scarce	Low	Fluvial and fluvio-marine environments. Remains of only coastal/estuarine organisms and infauna are preserved in an otherwise tectonic–fluvial dominated depocenters. Toward top, marine influence increases
Coniacian		Anaipadi S.St.	*Dravidosaurus blanfordi, Rhynchonella. Marginotruncana, Kosmaticeras* gr. *theobaldianum, Kosmaticeras theobaldianum crassicostata, Puzosia* sp., *Damesites* aff. *Sugata, Exogyra* (Costagyra) *fausta, Lopha* (Actinostreon) *diluviana, Proplacenticeras tamulicum, Nautilus,* Bored wood, encrusting oysters, and mollusca	Abundant to high	Rich to very rich	Subtidal to relatively deeper environments and associated fauna, often preferred locales of nektobenthic accumulation as well as terrestrial/coastal wood clasts
		Grey S.St.	*Teredolites, Thalassinoides,* oyster, gastropod, and mollusca	Abundant to high	Rich to very rich	High energy subtidal to supratidal environments. Predomination of coastal communities
Turonian		Kulakkanattam S.St.	Abundant wood fragments encrusted by oysters, *Pinna, Testudines, Exogyra haliotoidea, Lewesicereas anapadense,* mollusca. *Skolithos, Diplocraterion, Ophiomorpha,* and *Thalassinoides*	Abundant to high	Rich to very rich	Shallow intertidal to below storm weather wave base environments and subtidal shell banks. Dynamic depocenter as a result of high-frequency RSL. It also favored mixed siliciclastic–carbonate deposition and accordingly the faunal composition

Stage	Group	Formation	Fauna	Abundance	Richness	Interpretation
Cenomanian	Karai	Odiyam Sandy clay	*Exogyra, Alectryonia, Pecten, Whiteinella Arahaeocretacea, Gyrochorte comosa, Thalassinoides suevicus, Ophiomorpha nodosa, Palaeophycus tubularis, Scolicia isp. Taenidium Serpentinum, Skolithos isp.., Arenicolites isp.. Rhynchostreon suborbiculatum, Pycnodonte vesiculosa, Shark, Mantelliceras Mantelli, Mortoniceras rostratum, Pseudaspidoceras footeanum, foraminifera, and serpulid, radiolaria*	Scarce to high	Low to rich	**A mixed shallow marine environmental setting, probably resulted by change in provenance, and tectonic stability**
		Gypsiferous clay	*Rotalipora Subticinensis, Tubulostrum, Praeglobotruncana Helvetica. Hedbergella. Preaglobotruncana, Whiteinella, Pycnodus* sp.. *Lycochupea menakiae. Thalassinoides suevicus. Palaeophycus tubularis, Planolites isp.. Chondrites isp. Rosselia isp.. Taenidium isp. Palaeophycus tubularis and Rosselia isp. Macaronichnus isp., Ophiomorpha isp., Ophiomorpha nodosa, O. borneensis, Thalassinoides suevicus, Exogyra (Costagyra) costata, Squalicorax Baharijensis, Platypterygius, Gladioserratus, Psychodus decurrens, Eucalycoceras pentagonum,* and *radiolaria*	Scarce to high	Low to rich	**Flooding of marine shelf, sediment influx highly influenced by seasonal fluctuations, dominated by humid climate, resulting in abundant siliciclastic supply. It also thwarted sustenance and proliferation of carbonate dominated environment and associated fauna. Faunal proliferation and dwindling were dependent on environmental parameters**
	Dalmiapuram	Kallakkudi Calcareous S.St.	*Rotalipora appenninica, Rotalipora cushmani, Exogyra, Alectryonia, Tetrabelus seclusus, Parahebolites, blanfordi, Neohibolites* sp., *Pycnodon* sp., Phyloceratid, echinoid, bryozoa, and *Platypterygius indicus*	Abundant to high	Rich to very rich	**Establishment of typical coastal marine siliciclastic shelf associated with luxuriant, warm, normal marine waters and adjacent relatively deeper marine environments**
		Olaipadi conglomerate	Reworked fauna namely: Belemnite, rudist, coral, and serpulid	Rare to scarce	Low to significant	**Destruction of reef environment and predomination of tectonic-gravity, fluvial influenced sediment supply, resulting in only recycled faunal remains**

(Continued)

Table 4.3 (Continued)

Age	Fm.	Mbr.	Faunal and Floral Composition	Population Density	Diversity	Ecospace Dynamics
Albian		Dalmiya biothermal L.St.	Red algae, coral, bryozoa, gastropod, bivalve, echinoid, ostracoda, foraminifera, sponge, *Anomalinoides, Gavelinella plummerae, Gyroidinoidesglobosa, Lenticulina, Melobesioideae, Melobesioideae, Lithophyllum alternicellum, Pseudoamphiroa propria Quadrimorphina, Rastellum* (Arcostrea) *carinata, and Ostrea sp.*	Abundant to high	Rich to very rich	Typical reef forming organisms and associated fauna and algae
		Varagupadi biostromal L.St.	Bivalve, rudist, coral, algae, foraminifera, ostracoda, bryozoa, echinoid, *Melobesioideae, Melobesioideae, Lithophyllum alternicellum and Pseudoamphiroa propria Turrilites costatus, Acanthoceras sp., Mammites conciliatus, Nautilus, huxleyanus, Parachaetetesas vapattii, Sporolithon sp., Lithothamnion sp., Lithophyllum sp., Pseudoamphiroapropria, Neomeriscretaceae, Salpingoporella verticelata, and Agardioliopsis cretaceae*	Abundant to high	Rich to very rich	Establishment of shallow, relatively stable, normal marine depocenters and development of reef-reef associated environments. Reef forming organisms and reef dwellers thrived
		Grey shale	Palynoflora, *H. planispira, Parachaetetes asvapattii, Sporolithon sp., Lithothamnion sp., Lithophyllum sp., Pseudoamphiroa propria, Neomeris cretaceae, Salpingoporella verticelata, Agardioliopsis cretaceae,* Ostracoda, bryozoa, and gastropoda	Abundant to high	Rich to very rich	Marine flooding and resultant creation of deeper oxygen poor environments located adjacent to highly productive regions. Very-high frequency oscillations of oxygen poor-normal conditions as a result of RSL fluctuations. The durations of normal conditions increase toward top. Very high abundance of planktic and recycled benthic fauna

Stage	Formation	Member	Fossils			Environment
Aptian	Sivaganga	Terani clay	*Ptilophyllum, Gymnoplites, Pascoeites, Microcachyidites, Cooksonites, Aequitriradites,* and inoceramid	Rare to scarce	Low to significant	**Coastal freshwater lakes, inundated as a result of marine incursion. Freshwater environment supported only microfauna, while the marine incursion resulted in accumulation of recycled bioclastic components**
		Kovandan-kuruchchi S.St.	Early Cretaceous palynoflora	Rare to scarce	Low to insignificant	**Creation of subaqueous fan deltaic environments dominated by siliciclastic deposition and turbid conditions, as a result of which, only drifted and or extra-basinal palynofloral remains are preserved**
Barremian		Basal Conglomerate	*Globigerina boteriveca*	Rare to scarce	Low to insignificant	**Initiation of basin and coastal marine deposition under very high-energy conditions; only deeper regions contain certain microfauna**

The list of species, genus, family and other names, is a compilation from field work, laboratory examination and selected previous publications including but are not limited to Egerton (1845), Blanford (1862), Stoliczka (1873), Matley (1895), Kossmat (1895), Matley (1929), Rama Rao (1932), Gowda (1964, 1966, 1967), Sastry et al. (1968, 1972), Banerji (1972), Mamgain et al. (1973), Paul (1973), Narayanan (1977), Chiplonkar and Tapaswi (1979), Yadagiri and Ayyasami (1979, 1989), Yadagiri et al. (1983), Ayyasamy and Das (1990), Ayyasamy (2006), Ayyasamy and Rao (1987), Kale and Phansalkar (1992a,b), Chandrasekaran and Ramkumar (1993), Chandrasekaran et al. (1993), Ramkumar and Chandrasekaran (1996), Hart et al. (1996), Ramkumar (1997a,b, 2008), Govindan and Ravindran (1996), Govindan et al. (1996), Kohring et al. (1996), Kale et al. (2000), Misra et al. (2009), Ramkumar et al. (2004a, 2005a, 2010a), Kale (2011), Ramkumar and Sathish (2009), Underwood et al. (2011), Bragina et al. (2013), Prasad et al. (2013), Rai et al. (2013), Paranjape et al. (2014), Sugantha et al. (2015), and Verma (2015). This list includes generic and common names of the taxa at species and genus level. This list is neither exhaustive nor indicative and is provided only for an indicative purpose. Population and diversity of individual members, formations, and chronological units are assessments based on the field and laboratory observations of the authors and previous publications. These are utilized for comparison in the light of different depositional units, and lithological types, and to elicit information on ecosystem dynamics and resultant population and diversity changes as a function of environmental, depositional, climatic, sea-level, and other geological scenarios. Plates 4.1–4.4 depict few of these fossils.

Plate 4.1 **A:** Ptillophyllum—*Late Aptian Terani Clay Member;* **B:** *Petrified tree trunk measuring 18 m long and ca. 2 m diameter—Middle Turonian Kulakkanattam S.St. Member;* **C:** *Petrified wood fragments of various sizes strewn in silty sand—Early Turonian Odiyam Sandy clay Member;* **D:** *Well preserved shark teeth in the Cenomanian Gypsiferous clay Member;* **E and F:** *Trace fossils (*Zoophycus*) in the Albian Biostromal limestone Member;* **G:** *Trace fossil found in the upper part of Santonian Varakuppai lithoclastic conglomerate Member;* **H:** *Colonial serpulids in the Middle-Late Campanian Varanavasi sandstone Member;* **I:** *Thick population of* Gryphea *in the Early Maastrichtian Srinivasapuram gryphean limestone Member.*

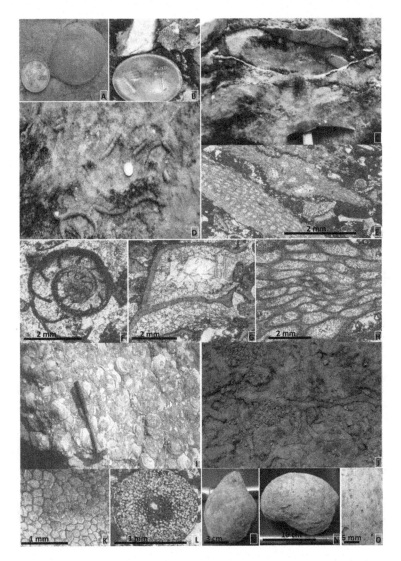

Plate 4.2 *A:* Stigmatophygus elatus—*Early Maastrichtian TANCEM biostromal limestone Member;* *B:* Fenestrate bryozoa—*Early Maastrichtian TANCEM biostromal limestone Member;* *C:* Meter-scale large clam in the Early Maastrichtian TANCEM biostromal limestone Member; *D:* Planolites Nicholson *found in rocks below storm deposits of the Early Maastrichtian TANCEM biostromal limestone Member. Occurrence of trace fossils below the storm deposits and their absence in the aftermath of storm (in beds above the storm deposits) indicates catastrophic devastation of natural habitat due to sudden sea-level change.* E: Orbitoid *found in the tidal channel deposits of the Early Maastrichtian TANCEM biostromal limestone Member;* F: *Cross section of foraminifera found in the Early Maastrichtian Kattupiringiyam Inoceramus limestone Member;* G and H: *Microphotographs of bryozoa—Early Maastrichtian Kattupiringiyam Inoceramus limestone Member;* I: *Thick population of* Inoceramus *in the Early Maastrichtian Kattupiringiyam Inoceramus limestone Member;* J: *Thick population of* Ophiomorpha irregulaire *occurring over the storm deposits of TANCEM biostromal limestone Member. Their occurrence and thick population suggest opportunistic colonization of ecospace vacated by native species in the aftermath of catastrophic event and sudden sea-level change;* K: *Microphotograph of* Inoceramus *shell—Early Maastrichtian Kattupiringiyam Inoceramus limestone Member;* L: *Microphotograph of echinoderm plate—Early Maastrichtian TANCEM biostromal limestone Member;* M: Terebratula—*Early Maastrichtian Srinivasapuram gryphean limestone Member;* N: Gryphea—*Early Maastrichtian Srinivasapuram gryphean limestone Member;* O: *Close-up view of the gryphean shell depicted in 'N' showing sponge boring.*

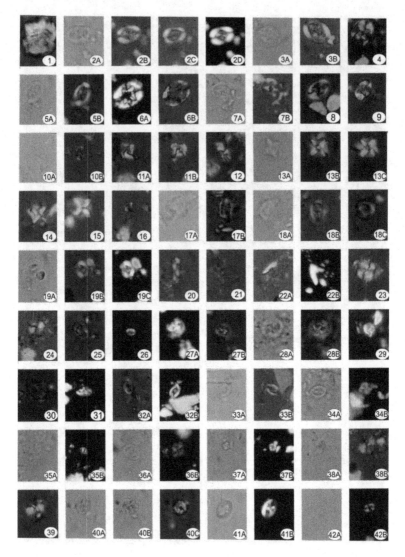

Plate 4.3 *1.* Braarudosphaera bigelowii *(Gran & Braarud) (Deflandre, 1947);* *2A–D, 3A and B.* Arkhangelskiella cymbiformis *(Vekshina, 1959);* *4.* Eiffellithus turriseiffeli *(Deflandre in Deflandre & Fert) (Reinhardt, 1965);* *5A and B.* A. cymbiformis *(Vekshina, 1959);* *6A and B, 7A and B.* E. parallelus *(Perch-Nielsen, 1973);* *8–9.* E. turriseiffeli *(Deflandre in Deflandre & Fert) (Reinhardt, 1965);* *10A and B.* E. gorkae *(Reinhardt, 1965);* *11A and B.* Micula swastika *(Stradner and Steinmetz, 1984);* *12.* M. murus *(Martini) Bukry, 1973;* *13A–C, 14–16.* M. decussata *(Perch-Nielsen, 1973);* *17A and B, 18A–C, 19A–C.* Stradneria crenulata *(Bramlette & Martini) (Nöel, 1970);* *20–21.* Microrhabdulus undosus *(Perch-Nielsen, 1973);* *22A and B.* Ceratolithoides kamptneri *(Bramlette and Martini, 1964);* *23–24.* Watznuaeria barnesae *(Black) (Perch-Nielsen, 1968);* *25.* Cribrosphaerella ehrenbergii *(Arkhangelsky) Deflandre in Piveteau (1952);* *26.* Cribrosphaerella *sp.; 27A–B, 28A–B.* Prediscosphaera cretacea *(Arkhangelsky) (Deflandre, 1968);* *29.* P. ponticula *(Bukry) (Perch-Nielsen, 1984);* *30.* P. spinosa *(Bramlette and Martini) (Gartner, 1968);* *31.* Prediscosphaera *sp.; 32A–B, 33A–B.* Zygodiscus spiralis *(Bramlette & Martini, 1964);* *34A–B.* Staurolithites crux *(Deflandre & Fert) (Caratini, 1963);* *35A–B.* Z. minimus *(Bukry, 1969);* *36A–B.* Ahmuellerella octoradiata *(Górka) (Reinhardt, 1966);* *37A–B.* Pseudomicula quadrata *Perch-Nielsen in Perch-Nielsen et al. (1978);* *38A and B.* Cyclagelosphaera deflandrei *(Manvit) (Roth, 1973);* *39.* W. barnesae *(Black) (Perch-Nielsen, 1968);* *40A–C.* Petrobrasiella? bownii *(Burnett, 1998);* *41A and B.* Chiastozygus litterarius *(Górka) (Manivit, 1971);* *42A and B.* A. regularis *(Górka) (Reinhardt and Górka, 1967). Source:* Rai, R., Ramkumar, M. and Sugantha, T., 2013 Calcareous nannofossils from the Ottakoil Formation, Cauvery Basin, South India: Implications on age, Biostratigraphic correlation and palaeobiogeography. In: Ramkumar, M., (ed.), On the sustenance of Earth's resources, Springer-Verlag, Heidelberg. 109–122; All forms are magnified to 2000 times.

*Plate 4.4 A: Close up view of the fossilized bone fragment of presumed dinosaur—Late Maastrichtian Kallamedu Formation; **B:** Closer view of a bone fragment found embedded in coarse grained sandstone of the Kallamedu Formation; **C:** Presumed egg casts of sauropod—Late Maastrichtian Kallamedu Formation; **D:** Closeup view of the egg cast; **E:** SEM image showing the oblique cross-sectional view of egg shell depicting the spherulite units separated by pores (indicated by arrows). The calcitization process has differential stability of the spherulitic units and pore fills in which the latter are often found to get lost; **F:** SEM photograph showing the cross sectional view of egg shell. Radiating spherulites of calcitic crystals are visible in the photograph. Black arrows at the bottom of the photograph indicate the nucleation sites; **G:** Nautilus in the Danian Anandavadi arenaceous Member; **H:** Brain coral fossil in the Danian Anandavadi arenaceous Member; **I:** (Source: Ramkumar, M., Anbarasu, K., Sugantha, T., Jyotsana Rai, Sathish, G. and Suresh, R., 2010a Occurrences of KTB exposures and dinosaur nesting site near Sendurai, India: An initial report. International Journal of Physical Sciences 22, 573–584; All the forms are to 2000 times). **1A–C, 15.** Discorhabdus ignotus (Górka, 1957; Perch-Nielsen, 1968); **2A and B.** Eprolithus floralis (Stradner, 1962; Stover, 1966); **3–4.** Pseudomicula quadrata Perch-Nielsen in Perch-Nielsen et al. (1978); **5A and B.** Holococcolith sp.; **6.** Prediscosphaera sp.; **7.** Uniplanarius sp.; **8A and B.** Petrobrasiella? bownii (Burnett, 1998b); **9.** Eprolithus sp.; **10.** Corollithion exiguum (Stradner, 1961); **11A and B.** Ottavianus terrazetus (Risatti, 1973); **12A and B.** Lucianorhabdus cf. L. maleformis (Reinhardt, 1966); **13A and B.** Calculites obscurus (Deflandre, 1959) Prins and Sissingh in Sissingh (1977); **14A–C.** Ceratolithoides pricei (Burnett, 1998a); **16, 17A and B.** Biantholithus cf. B. sparsus; **J:** Tabulate coral colony in the Danian Anandavadi arenaceous member; **K:** Bivalve shell casts in the Danian Anandavadi arenaceous Member.*

CHAPTER 5

Depositional History and Habitat Heterogeneity

The sequence stratigraphic concepts have unequivocally established that the loci of sedimentation and the pattern of facies stacking are controlled by tectonics−climate-relative sea-level fluctuation−sediment influx/production (Vail et al., 1977; Sarg, 1988; Catuneanu, 2006; Catuneanu et al., 2011). Taking cues from these and given cognizance to the nexus between sea-level fluctuations, depositional systems (Holland, 2012; Holland and Christie, 2013; McClure and Lockwood, 2015) and resultant facies type occurrence and distribution (Smith and Benson, 2013; McMullen et al., 2014; Ritterbush et al., 2014), the relative roles of geological events including sea-level fluctuations over habitat diversity, area of distribution, and stratigraphic heterogeneity (Darroch, 2012; Zuschin et al., 2014) are examined in this chapter. This chapter also examines whether documentation of facies types (Figs. 4.2, 4.6−4.8; Table 4.2), and geological events namely, tectonic, and relative sea-level changes (Fig. 4.3), among others, could help document habitat dynamics and associated biotic occurrence, population, and diversity (Table 4.3; Fig. 4.2).

5.1 TECTONIC INFLUENCE ON HABITAT DYNAMICS

Tectonic events exercise first-order control over landscape evolution. Their intensity and region of influence control the ecospace availability, creation, destruction, shifting, and alteration and, in turn, biodiversity. Few studies have documented the direct and indirect influence of tectonics over biodiversity-habitat dynamics (Badgley and Finarelli, 2013; Domingo et al., 2014; Mihaljević et al., 2014; Villafaña and Rivadeneira, 2014; Zuschin et al., 2014). Enumeration of the tectonic structures and depositional history of the Cauvery Basin indicated that the initial block faulting and inception of sedimentation occurred during Late Jurassic-Early Cretaceous that paved way for the development of marine and associated onland and other ecospaces over former continental regions. Thenceforth, the intensity of tectonic control over

Eustasy, High-Frequency Sea-Level Cycles and Habitat Heterogeneity.
DOI: http://dx.doi.org/10.1016/B978-0-12-812720-9.00005-X

sedimentation and habitat development was diminutive (Prabakar and Zutchi, 1993; Ramkumar, 1996; Ramkumar, 2015) with an exception of Middle Santonian event that had brought extensive regions under the influence of marine forces (Ramkumar et al., 2005a). The Cenomanian tectonic activity that eroded the Dalmia biohermal deposit was more of reorganization of existent marine habitat. Other than these, the depositional history of the Cauvery Basin remained tectonically quiescent (Ramkumar and Berner, 2015).

Having established the limited and episodic role of prevalent tectonic events during the evolutionary history of the Cauvery Basin, the contribution of other events such as relative sea-level fluctuations (eustasy and local–regional high-frequency changes) over the depositional pattern, facies distribution, and habitats is examined.

5.2 DEPOSITIONAL HISTORY AND HABITAT DYNAMICS

Any sedimentary environment can be defined by physical, chemical, and biological conditions in a geomorphic set up (Reineck and Singh, 1980). Each sedimentary environment or environmental conditions produce a distinct facies, recognizable through a set of sedimentary, tectonic, textural, mineralogical, and biological characteristics (Ramkumar, 1999, 2015). Idealistic profiles of tectonogeomorphology and depositional systems/sedimentary environments from source-sink are presented in Figs. 1.1 and 1.2. Fig. 1.2 presents the depositional systems and plausible sedimentary facies types that can be generally expected in each of the system. As depicted in this figure, changes in relative bathymetry/sea-level, as a result of tectonics–climate-relative sea–level sediment influx/production impact these depositional systems in terms of variations in stratigraphic occurrences of facies types. As explained in previous chapters, the changes in environmental parameters and habitat characteristics influence the biotic realm. Thus, a systematic documentation of facies types and stacking pattern might reveal the prevalent habitat dynamics. Explicit or implicit application of this premise is found in fewer studies (Holland and Zaffos, 2011; Darroch, 2012; Smith and Benson, 2013; McMullen et al., 2014; Ritterbush et al., 2014; Zuschin et al., 2014). Taking cognizance of this premise, this section presents a brief description of prevalent habitat changes in terms of creation, destruction, alteration, reduction, expansion, and shift, as explicit in facies types and their stratigraphic variation.

As a result of initial rifting and associated down-throw of the Pondicherry subbasin, widespread transgression took place and initiated the sedimentation of the Sivaganga Formation. Significant sedimentation commenced with the establishment of a fluvial source onland and a submarine fan delta in the basin (represented by the Kovandankurichchi Sandstone Member). Gradation of this predominantly sandy member into deep marine claystone—siltstones (Terani Member) indicates increase in bathymetry of the depocenter. It was brought to an end due to renewed faulting and introduced an angular unconformity, intensive erosion, and resedimentation of older sedimentary rocks. The rejuvenated sedimentation was through the deposition of shale and shale—limestone cyclic beds of the Grey Shale Member. The depocenter was partially and periodically closed. The water column was also partially/periodically stratified. The grey shale beds were deposited during these closed and stratified conditions. The bioclastic limestone bands were deposited when open marine and well-circulated conditions prevailed. A general increase in thicknesses of the limestone interbeds toward top is recognizable. It indicates an increase in the duration of the openness of the sea, which culminated in the development of a biostromal member. In addition to reef-dwelling microfauna, this biostromal member contains coral clasts and algal fragments with varying proportions of siliciclastics, clasts of bryozoa, bivalvia and gastropoda, signifying deposition in subtidal to storm weather wave base regions with photic conditions. Typical coral bioherms developed over this member that moved gradually toward offshore regions owing to a fall in the sea-level. At top of this biohermal limestone, a major erosional surface associated with faulting and regression is observed. This faulting had exposed the subtidal—storm weather wave base deposits to subaerial conditions which led to karstification. It also led to the deposition of the Olaipadi Conglomerate Member which contains large boulders (many of which are more than 10 m in diameter) of basement rocks, lithoclasts of similar size, that were drawn from underlying bioclastic and coral limestone, Terani claystone, and lithoclasts of older sedimentary conglomerates. These clasts are embedded in basinal clay sediments signifying syndepositional tectonics and recycling of sediments and basement rocks. Deposition of argillaceous sediments ceased by rejuvenation of influx of coarse to finer clastics and suspended sediment load from the fluvial source. This led to the deposition of the Kallakkudi Calcareous Sandstone Member. This member is sandy in the southern region and clayey in the northern region, indicative of the depositional topographic slope and distance from provenance. The recurrent Bouma sequences each toping with a gypsiferous layer, followed by an erosional surface in this member indicates the dynamic nature of depocenter, episodic closure of the sea, exposure of sediments to subaerial conditions, rejuvenation of deposition with a rise in the sea-level, deposition under the influence of turbidity currents, and gradual facies change from near shore to deep sea. Deposition of this member took place in a slowly sinking basin. Conversely, deposition with episodic sea-level rises and falls coupled with active fault block adjustment (to a minor degree) after major movement could also explain the facies characteristics and stacking pattern.

With the sinking of the coastal basin and/or sea-level rise, deep-marine conditions were established and thick pile of the Karai Formation clay was deposited. Deposition of about 450m-thick succession comprising clay alternated with ferruginous silty clays and gypsiferous layers suggests a well-developed fluvial system onland which persistently supplied suspended sediment load to far offshore regions. Thick population of belemnites, silty admixture, alternate thin-thick laminae of ferruginous silty clay, and gypsiferous clay bands are frequent in the southern region, which is indicative of the deposition in shallower region of paleosea also. This shallower region was periodically exposed subaerially due to minor sea-level oscillations to produce evaporites. The top surface is marked by a pronounced erosional surface, which suggests major regression. This erosional surface is overlain immediately by subtidal−supratidal ferruginous sandstones along with shell banks typical of an estuary and shell hash typical of shallow water shoals/distributary mouth bars that represent the Kulakkanattam and Grey Sandstone members of the Garudamangalam Formation. Together, their occurrences indicate shoreline retreat and advancement of fluvial system over the former offshore areas. This inference is substantiated by the sudden appearance of large tree trunks in these sandstones. At places, pebble−gravel-sized petrified wood clasts are extremely abundant, suggestive of extensive sprawl of terrestrial and coastal forest habitats that experienced severe erosion and destruction of the marine habitat due to sea-level lowering and fluvial incision. Deposition of these sandstones in estuary or tidal flat, beach and distributary mouth bar is indicated by excellently preserved tidal couplets akin to the present day coastal deltaic systems (Ramkumar, 2000, 2003; Ramkumar et al., 2000a,b). Relatively deep-water conditions were restored again in this part of the basin with the introduction of the deposition of the Anaipadi Sandstone Member. A break in sedimentation, probably influenced by major regression, was witnessed at the end of the Anaipadi Member.

Renewed transgression during the Middle Santonian associated with widespread erosion of basement rocks and older sedimentary rocks and resedimentation of them in the newly created depocenters. The lowermost member of the Sillakkudi Formation is a fluvial unit and shows gradual transition to marine influence toward top. The continued sea-level rise submerged the fluvial/estuarine mouth and a typical coastal marine member in subtidal to intertidal environments started to develop. It was followed toward top by the deposition of the Varanavasi Member, whose deposition took place in wide shelf. Frequent occurrences of pebbly sandstone layers, erosional surfaces, reworked fauna, and localized serpulid colonies found at top of the Varanavasi Member indicate cessation of sediment supply, sea-level lowering, reduced circulation, and lower energy conditions. The renewed transgression during the Late Campanian—Early Maastrichtian was marked by widespread erosion of basement rocks and older sedimentary deposits. The resultant Kallankurichchi Formation is essentially a bioproduced carbonate unit that denotes the cessation of a supply of fluvial sediment which existed during Santonian—Campanian. As the initial marine flooding started to wane, gryphean colonies started proliferating marking the beginning of carbonate sedimentation. As the sea-level gradually increased, the gryphean bank shifted toward shallower regions and the locations pre-viously occupied by coastal conglomerates became middle shelves, on which inoceramid limestone started developing. Break in the sedimen-tation of this member was associated with a regression, which trans-formed the middle—outer shelf regions into intertidal—fair weather wave base regions. These newer depositional conditions resulted in the erosion of shell banks and middle shelf deposits and their redeposition into biostromal deposits (Tancem Biostromal Member). Again, the sea-level rose to create a marine flooding surface. As a result, gryphean shell banks started developing more widely than before, forming the Srinivasapuram Gryphean Limestone Member. The occurrence of a nondepositional surface at top of this member and deposition of shallow marine siliciclastics (Ottakoil Formation) in a restricted region immediately over the predominantly carbonate depocenter and conformable offlap of much younger fluvial sand deposits (Kallamedu Formation) are all suggestive of a gradual regression associated with the reestablishment of a fluvial system at the end of the Cretaceous Period.

Occurrence of well-sorted calcareous sandstones with frequent upward fining sequences, large-scale cross bedding, abundant but patchy occurrences of *Stigmatophygus elatus* and extensive occurrences of ichnofauna together with nannofossil evidences (Rai et al., 2013) suggest deposition of the Ottakoil Formation in shallow, open, low energy, normal saline characteristics in a near shore/marginal marine regime that experienced low sedimentation rate under the influence of seasonal and or episodic increase in energy conditions. The Kallamedu Formation was deposited under coastal plain environment that was periodically inundated by fresh water overflown from ephemeral shallow river channels (Ramkumar et al., 2010a). Its facies characteristics indicate deposition in river channel, flood plain, and overbank environments of fluvial system and periodically open/closed coastal lagoon located adjoining coastal region that was under the seasonal influence of marine and fluvial agents (Ramkumar et al., 2013). Toward top of the Kallamedu Formation, paleosols are recorded, implying abandonment of the river system and restoration of continental conditions at the end of Cretaceous (Sugantha et al., 2015). At the beginning of Danian, transgression occurred, which covered only the eastern part of the Kallamedu Formation. Presence of a conformable contact between the Anandavadi Member and the Kallamedu Formation and initiation of carbonate deposition from the beginning of Danian are indicative of absence of fluvial sediment supply and tectonic activity at this time. Steady rise of sea-level and establishment of a shallow, wide shelf with open circulation paved way for the deposition of the Periyakurichchi Member. At top, this member has distinct erosional unconformity, which in turn, when interpreted along with the presence of a huge thickness of continental sandstone (>4000 m thick Cuddalore Sandstone Formation of Miocene to Pliocene age), clearly indicates the restoration of continental conditions in this basin. Absence of any other marine strata over the Cuddalore Sandstone Formation suggests that the sea regressed at the end of Danian and has never returned to this part of the basin.

Thus, as described in this section and summarized in Figs. 4.3, 4.6–4.8, habitats were changing as a function of sea-level fluctuations and associated climatic reversals during Barremian–Danian in the Cauvery Basin, with subordinate, but highly dramatic roles played by tectonics during basin formation, Cenomanian and Santonian. The facies characteristics, their distribution, stacking pattern, and contact relationships have expressed their coupled nature with changing environmental parameters, promoted by these sea-level fluctuations. The Cauvery Basin strata record all the six global sea-level peaks separated by type 1 sequence boundaries and other correlative surfaces coeval with third-order cycles of sea-level (Ramkumar et al., 2011) within which many high-frequency sea-level cycles are perceptible (Fig. 4.3). In addition, as explained in the previous section, there were minor tectonic events that were coeval with few of the sea-level fluctuations and resultant bathymetric changes in the depositional systems. Hence, it is imperative to examine the scale of the sea-level fluctuations and their genetic (eustasy or otherwise) causes.

CHAPTER 6

Proxies for Habitat Heterogeneity and Dynamics

Establishment of the dominant role played by sea-level fluctuations over facies distribution, stratal patterns, and cycles-in-cycle nature of depositional systems and resultant rock records, which in turn served as habitats to biota either positively or negatively, necessitates examination of dependence of facies characteristics on climatic and environmental conditions. The extent of dependence and the possibility of elucidation of this trait through available tools also need to be analyzed. These could help examination of facies characteristics, facies succession, and stacking pattern as proxies for assessment of habitat heterogeneity and dynamics. This approach may supplement the conventional assessment methods of habitats based on occurrence, population, and diversity of biota, which are often subjective, tenuous, and discordant and suffer from various biases (Ramkumar, 2015a,b).

6.1 GEOCHEMICAL INDICATORS

The stratigraphic record is the outcome of an exogenic system consisting of geologic setting, changes in sea-level, changes in geochemical reactions between the sea and the earth, climate and processes of sediment formation (Srinivasan, 1989), and diagenesis (Ramkumar, 1999). The ensuing sedimentary records show differences in bulk chemistry as a result of changes in different combinations of minerals, either bioproduced or brought into the depocenters and neoformed during post depositional times, all of which were influenced either partly or wholly by the climatic conditions (Ramkumar, 2015a,b). According to Veizer et al. (1997), the sedimentation system is dominated by cyclic processes that operate on a hierarchy of temporal and spatial scales on which short-lived events are superimposed. An ability to recognize these through the analysis of stratigraphic variation of geochemical composition of strata,

Eustasy, High-Frequency Sea-Level Cycles and Habitat Heterogeneity.
DOI: http://dx.doi.org/10.1016/B978-0-12-812720-9.00006-1

a la., chemostratigraphy, has emerged to be a reliable tool (Ramkumar, 1999, 2015c,d) even where other conventional methods fail or show limitations (Ramkumar et al., 2011, 2015d). The general knowledge about a geochemical system allows establishing a finite number of processes governing the sedimentary system namely, detrital input, changes in provenance and quantum of sediment influx and climate, and others (Montero-Serrano et al., 2010). Given cognizance to all these, and the sensitivity of geochemistry toward tectonics, climate, sea-level fluctuation, and sediment influx/production (Srinivasan, 1989; Ramkumar, 1999, 2001, 2015c,d; Ramkumar and Berner, 2015; Ramkumar et al., 2004b, 2005b, 2006, 2010c, 2011), few selected geochemical profiles of Barremian–Danian strata of the Cauvery Basin (Fig. 6.1) were examined and supplemented with previous studies (Ramkumar, 2015c; Ramkumar and Berner, 2015; Ramkumar et al., 2004b, 2005b, 2006, 2010b,c, 2011), prevalent sea-level fluctuations (Fig. 4.3), tectonic, and other geological events (Figs. 4.1 and 4.3) to assess the utility of facies as visualized through chemostratigraphy as a tool for documenting habitat heterogeneity and dynamics and also the spatial and temporal scale of inferred heterogeneity.

6.2 CHEMOSTRATIGRAPHY AS A PROXY

Examination of the selected geochemical profiles of the Barremian–Danian strata of the Cauvery Basin reveals the presence of two second-order cycles separated by Santonian (as explicit in the polynomial curve—Fig. 6.1), implying major change in the depositional pattern-prevalent environment-facies succession in the basin. The temporal resolution pattern of the element Zr, among others, is more explicit. As Zr undergoes little diagenetic alterations (Andrew et al., 1996; Whitford et al., 1996; Das, 1997), its short and prominent peaks exactly coinciding faulting events (as inferred independently in the field structures and contact relationships of beds— Ramkumar et al., 2004a) and associated change in sedimentation pattern as reflected in lithofacies characteristics (Table 4.2; Figs. 4.1 and 4.2) suggest influx of Zr immediately after tectonic events (Ramkumar and Berner, 2015).

Notwithstanding the tectonic influence on depositional pattern and resultant change across Santonian, within these two second order cycles, the geochemical profiles are divisible into six major

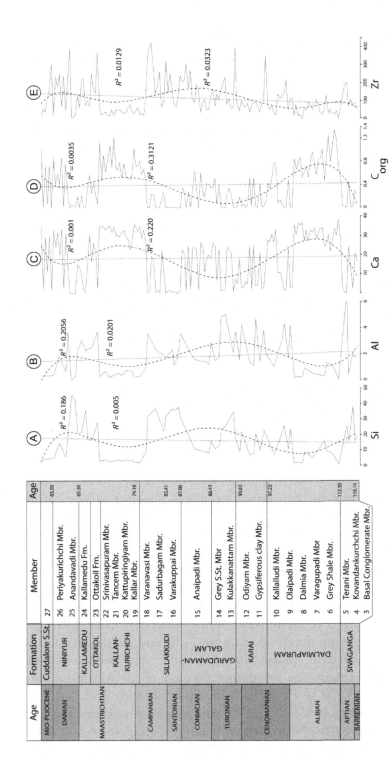

Figure 6.1 *Stratigraphic variations of elemental concentrations and organic carbon.* Si, Al, Ca, and C_{org} are in weight percentage. Zr is in ppm. Solid line curve is indicative of absolute values while the dotted line (.....) indicates linear trend and the dashed line (— —) indicates polynomial trend. Variations are shown against the chronostratigraphic and lithostratigraphic subdivisions and absolute age.

chemozones (Ramkumar et al., 2010b, 2011). These chemozones are synchronized with the six third-order relative sea-level cycles (compare the geochemical profiles depicted in Fig. 6.1 with sea-level curve depicted in Fig. 4.3). There are many high-frequency cycles and chemozones stacked within these third-order sea-level cycles-chemozones. The lithofacies characteristics (Table 4.2) along with the relative sea-level variations (Fig. 4.3) demonstrate the prevalent high-frequency variations of sea-level and resultant alternate depositions of siliciclastics and carbonates in the Cauvery Basin.

The large-scale facies alternations are either sympathetically or antisympathetically followed by the absolute concentration curves of Si–Al (Fig. 6.1A and B) and Ca (Fig. 6.1C). Decrease in Si and many metals typical of heavy minerals are interpreted as the result of transgressions (Hild and Brumsack, 1998). Conversely, the reduction of carbonate and C_{Org} contents (Fig. 6.1D) are found to be associated with regressions. The occurrences of inverse relationships between Si–Al–Zr (Fig. 6.1A, B, and E) and Ca–C_{Org} (Fig. 6.1C and D), lithofacies alternations between siliciclastics and carbonates (Tables 4.1, 4.2) and their synchronicities with sea-level lows and highs, respectively (Fig. 4.3) allow interpretation of sea-level controlled depositional pattern in this basin. As could be observed elsewhere (Rachold and Brumsack, 2001; Hofmann et al., 2001; Boulila et al., 2010), whenever siliciclastic deposition ceased, carbonate deposition was initiated (Warzeski et al., 1996) in this basin. Erosion during the periods of lower sea-level is also influenced by the proximity to source rocks. This inference is supported by the occurrences of paleochannel courses (Fig. 4.1) and their association with siliciclastic deposits (Figs. 4.2 and 4.7). Occurrences of unaltered lithoclasts and feldspar clasts in rocks that immediately follow regressive surfaces also suggest the prevalent mechanical erosion, rapid, and short duration of transport and quick burial. Such rapid, physical erosion and textural immaturity of ensued sediments might have produced covariation of Si and Al and other elements associated with quartz, feldspar, and other silicates. This interpretation explains the sympathetic nature of Si–Al during sea-level lowstands. These lithofacies alternations and inverse relationships between Si–Al and Ca can also be considered to have been influenced by the bathymetric variations, triggered by climatic reversals (Srinivasan, 1989; Fenner, 2001a; Pearce et al., 2003) and resulted in modification of prevalent habitats.

6.3 GEOCHEMICAL PROXIES OF CLIMATE, SEA-LEVEL, SEDIMENT INFLUX

The roles of climatic reversals on sediment influx, sea-level fluctuations, and finally over the facies succession and stacking pattern are expressed through the geochemical profiles of the Cauvery Basin. For example, the pre-Albian deposits are predominantly continental/coastal marine in nature and indicate humid climatic conditions (Wortmann et al., 2004), accompanied by chemical weathering. The warm-arid conditions and physical weathering were the norms during the deposition of the Olaipadi Member (Late Cenomanian-Early Turonian). Based on the configuration of the Cauvery Basin, and prevalent proximity between the provenance and depositional sites, it is perceivable that the modifying factors of lithological composition might have had little or no impact (Spalletti et al., 2008). If deposition under dry conditions is assumed, the climate may not have exercised control over chemical weathering. Thus, the chemical immaturity of the sediments of the Cauvery Basin might be the result of a combination of factors, including, but not limited to, tectonics, basin configuration, proximity between source and depositional site, and general warm-arid climatic reversals/cycles. The abundance of Zr in clastic rocks is considered to be the result of detrital influx as well as sediment recycling (Spalletti et al., 2008). Zirconium is mostly concentrated in zircons that accumulates during sedimentation while less resistant phases are preferentially destroyed (Alvarez and Roser, 2007), and hence, deposition during sediment starvation results in enhanced Zr content. This trait helps interpreting stratigraphic variations of Zr as the result of changes in weathering conditions in the hinterland or change of source regions (Bellanca et al., 2002). Zr is also considered to be effective in discriminating sediments of different tectonic histories (Andreozzi et al., 1997). Occurrences of generally higher levels of Zr (Fig. 6.1E) all through the Barremian–Danian with the exception of Late Campanian to Middle Maastrichtian (Kallankurichchi Formation) and its many episodes of positive excursions over this background value suggest sediment starved nature of the basin (meaning subdued levels of chemical weathering) and significant recycling of older sedimentary rocks. Peak enrichments of Zr during Middle Cenomanian (Gypsiferous Clay Member), Late Cenomanian (Odium Member), Middle Campanian (Varakuppai Member), Middle-Late Maastrichtian (Ottakoil Formation), and Early Danian (Anandavadi Member) indicate the influx and cessation of terrigenous materials which in turn was controlled by variations in source area weathering and/or a change from more humid to more arid conditions or tectonic movements (Munnecke and Westphal, 2004). Sandulli and Raspini (2004) interpreted the elemental cycles that occur in tune with unique facies associations as the results of precession and obliquity periodicities, and the facies bundles and super bundles to be the results of short and long eccentricity cycles. Similar inferences could be made to the rocks under study.

Based on stable isotopic analyses, Zakharov et al. (2006) computed paleotemperatures from 16.3 to 18.5°C for Albian and Turonian, and 19.8−21.2° for Early Maastrichtian in the Cauvery Basin. At this time, India drifted from warm-temperature climate zone into the nearby sub-tropical zone with the warm and perhaps humid climate. A global sea-level low at the Aptian−Albian boundary is recognizable at about 112 Ma (Hancock, 2001). Similar to the platform crisis observed by Schulze et al. (2004) during Middle Cenomanian in the west-central Jordan, reduction of carbonate, increase of clay, and organic matter in rock records coincident within sea-level rise are observed in the Cauvery Basin. However, typical grey shale with highest C_{Org} (Fig. 6.1D) and grey sandstone with higher C_{Org} occur during Early Albian and Late Turonian in this basin. The Early Albian event is represented by alternate beds of grey shale and grey bioclastic limestone among which the thickness of limestone progressively thicken toward top with concomitant reduction of thickness of shale beds (Ramkumar et al., 2004a). It is similar to the Middle Cretaceous Event beds described by Coccioni and Galeotti (2003) from the Scaglia Bianca Formation of Umbria−Marche Basin. While the increase of carbonate content and thicknesses of limestone beds may be associated with productivity cycles, the clay rich beds may represent dilution cycles (Böhm et al., 2003; Boulila et al., 2010), and thus, a general increase in productivity and concomitant reduction of dilution cycles could be inferred from this phenomenon. Keller et al. (2004) observed cyclic limestone/shale deposition of the lower Bridge Creek Limestone Member of the Pueblo section and interpreted them as a product of orbital cycles.

Anderson et al. (2001) concluded that the ratio of carbon-to-phosphorus in sediments should be used to assess the biogeochemical behavior and primary productivity through geologic time. Generally, C_{Org} in sediments is derived from the remains of aquatic and terrigenous plants, which are degraded during their transport to the site of deposition and during early diagenesis. Hence, preservation of organic matter in sedimentary records depends on environmental conditions, and thus, the coupling between organic matter preservation and depositional environment can be used as a key to deduce paleoenvironmental conditions (Littke et al., 1998). Jendrzejewski et al. (2001) commented that the production, accumulation, and preservation of organic matter are very sensitive to oceanographic and climatic changes. Tu et al. (1999) and Pellenard et al. (2014) have demonstrated the nexus between marine organic matter, atmospheric CO_2 content, climate, and sea-level. The sea-level rise and increase in shelf area are believed to be mechanisms for increased productivity and burial of organic matter (Hofmann et al., 2001; Pearce et al., 2003). Oxic nature of the depositional waters during the sea-level lowstands and associated deposition of siliciclastics thwart accumulation and preservation of organic carbon in the sediments (Friedman and Chakraborty, 1997). This trait effectively restricts higher concentration of organic matter associated only with carbonates (Fenner, 2001a). As these conditions can prevail during higher sea-level conditions (Fenner, 2001b), accumulation and preservation organic matter occur during warm climatic conditions and sea-level highstands. Jendrzejewski et al. (2001) and Fenner (2001a) reported maximum flooding surfaces with enrichments of total organic content in Albian sediments. Occurrences of alternation of siliciclastics and carbonates, among which higher C_{Org} in carbonates coupled with record of global signals of sea-level highs in the Cauvery Basin, help interpret falling pCO_2 due to the organic matter burial, resulting in climatic cooling (Jeans et al., 1991; Uličny, 1992; Raymo et al., 1997) and sea-level decrease (Tu et al., 1999). In due course, this decrease might have promoted shelf and continental erosion, siliciclastic influx, and oxic waters and reversed the cooling trend, which explain the lithofacies alternations and association of C_{Org} with carbonates. The C_{Org} (Fig. 6.1D) apparently mimics Ca (Fig. 6.1C), explaining its dependence on carbonate deposition-climate-sea-level (Boulila et al., 2010). Low C_{Org} except during the Albian, Latest Campanian-Middle-Late

Maastrichtian and Late Danian are consistent with bioturbation, benthic organisms, and oxygenated waters of deposition. Dependence of C_{Org} availability with carbonate deposition indicates its marine biological origin (Peryt and Wyrwicka, 1993) and low bottom water oxygenation (Pratt, 1984). Predomination of marine organic matter and the inferred absence of detrital organic matter may indicate the absence of significant continental vegetation. However, occurrence of less than 0.4% in most of the profile could be the result of poor preservation conditions (Jendrzejewski et al., 2001) and also due to lower influx of terrestrial organic matter. This inference is corroborated by the antisympathetic nature of Al and C_{Org} (Fig. 6.1B and D).

During the Cretaceous Period, the Indian plate was surrounded by seawater akin to the present day Australia. Owing to the prevalent greenhouse effect, the temperature gradient between the tropic and polar regions was low. Due to this, climatic cooling and resultant sea-level fall might have lengthened fluvial systems dramatically and transported detrital material. This dramatic shift in the sediment influx pattern might have balanced C_{Org} burial through cessation of carbonate deposition. Influx of continental waters coupled with reduced bathymetry significantly increase available oxygen in the waters, increase the mixing (Kampschulte et al., 2001; Jendrzejewski et al., 2001), and promote the destruction of C_{Org} (McKirdy et al., 2001; Fenner, 2001a). This mechanism may explain the alternate occurrence of siliciclastic and carbonate deposits as a consequence of sea-level fluctuations in this basin. Based on the occurrences of all the six global sea-level peaks, recognized independently through foraminifer (Raju and Ravindran, 1990; Raju et al., 1993), bathymetric estimates based on lithofacies succession (Ramkumar et al., 2004) and geochemical data (Ramkumar et al., 2005b, 2010a,b,c, 2011) influence of eustatic sea-level changes, global-scale climatic fluctuations over the stratal pattern, and facies succession, and thus, dynamics of depositional systems are affirmed. By implication, influence of environmental parameters, oceanographic processes, and climatic factors over habitat dynamics similar to the facies distribution is affirmed. It also demonstrates that unique habitats can be recognized through geochemical proxies similar to the recognition of facies types and facies associations as individual habitats.

6.4 FACIES AS PROXY FOR HABITAT HETEROGENEITY AND DYNAMICS

From the previous sections, it is demonstrated that the sea-level fluctuations have intimate coupling with the facies characteristics, and their spatial and temporal distribution, as a cumulative response to climatic, tectonic, and environmental parameters prevalent. The facies succession follows a cyclic pattern interspersed with perturbations. Notwithstanding the perturbations, the distinctness of the cycles-in-cycle is recognizable in facies and other criteria. Ramkumar et al. (2010b) recognized second, third, and higher-order chemozones of the Cauvery Basin to be statistically distinct, meaning, the chemozones to be significantly different from each other. It supports the use of facies characteristics and facies succession as proxies for assessing habitat heterogeneity and dynamics. As demonstrated in previous sections, unique facies types and succession are unique habitats that can be tracked by geochemical proxies as well.

CHAPTER 7

Implications and Future Trend

7.1 BIOTIC-HABITAT HETEROGENEITY OF THE CAUVERY BASIN

The Barremian−Danian strata of the Cauvery Basin were deposited under the influence of eustatic high-frequency sea-level fluctuations within six third-order cycles, which in turn were embedded on two second-order cycles. The high-frequency and third-order cycles were of exclusively eustatic in origin, However, few of the third-order cycles were partly influenced by basin scale tectonics. While the pre-Santonian is characterized by frequent environmental changes and texturally immature mixed siliciclastics and argillites, the younger deposits are characterized by relatively stable depocenters, carbonate deposition, textural inversion, and maturity. At each major unconformity, basement rocks and older sedimentary rocks were eroded and transported to newer depocenters to deposit lithoclastic conglomerates. A major tectonic movement during Santonian that caused transgression in areas that remained topographically highlands since inception of the basin (Barremian) had changed the depositional pattern from dynamic depocenters to more stable shelf conditions (Ramkumar et al., 2005a). Except this, the depositional history of this basin remained primarily under the influence of sea-level variations.

Though it is in common knowledge that the relative sea-level fluctuations, particularly the short-term, high-frequency cycles affect the habitats, environmental niches, and thereby the occurrence, distribution, population, and diversity of the lives that thrive in the zone of sea-level fluctuation, no systematic study had ever been conducted to document and understand the influences and impacts. The present study had revealed that the eustatic and tectonoeustatic cycles were instrumental in creation, modification, and destruction of habitats and are aptly reflected in the occurrence, abundance, and species diversity.

Eustasy, High-Frequency Sea-Level Cycles and Habitat Heterogeneity.
DOI: http://dx.doi.org/10.1016/B978-0-12-812720-9.00007-3

While the carbonates contain significant biodiversity and abundant population, the siliciclastics have lesser diversity and abundance. It may be due to two factors, namely the lithological bias as the carbonates have abundant bioproducers as well as get cemented syndepositionally, thwarting late stage diagetic destruction and also due to the fact that, the periods of siliciclastic deposition bring in acidic/less saline waters and introduced instability of salinity gradients in the habitats, which may not have been tolerated by the biota. These conditions might have resulted lesser abundance and diversity in siliciclastic lithologies.

This lithologic bias has got an unintended consequence on the traditional view on the depositional environments of the Cauvery Basin; i.e., owing to the predominant occurrences of highly diverse and thick populations of holomarine fauna and flora, *almost* all the previous workers in this area (except few that studied the Kallamedu Formation and the lower part of the Sillakkudi Formation), presumed purely *marine* conditions of deposition in the entire basinal history. However, elucidation of the basin fill history in the light of habitat heterogeneity has led to the proposition that though major part of depositional history of this basin took place under the influence of varying fluvial sediment supply, the prevalent terrestrial, coastal lowland, coastal swamp, onland fluvial channel, estuary, tidalflat, beach, lagoon, and supratidal habitats were *equally existent* and produced thick and widespread sedimentary facies types as were the shelfal and bathyl environments, but were largely overlooked due to their (fluvial and fluvio-marine deposits) relatively lesser fossiliferous nature. Owing to the fossil abundance, the holomarine deposits earned exclusive marine signature to the Cretaceous deposits of the study area, while the other deposits sandwiched between these marine deposits were relegated to oblivion and were presumed to be literally nonexistent or of minor importance. This is startling and the fact has come to light only due to the appreciation of habitat heterogeneity.

Though the nonholomarine habitats supported life, as evident from ichnofaunal assemblages and by drawing analogy with modern examples, prevalence of either very high, very frequently torrential energy conditions (as evidenced by large boulder—gravel imbricate lithoclasts at every unconformity all through the depositional history), or very high turbidity (as evidenced by thick claystone–siltstone characteristics of many of the siliciclastic deposits such as the Karai Formation) or prevalence of highly unstable niches that were under the influence of tidal and storm depositional settings (such as the Garudamangalam Formation; Habitat heterogeneity over geographic space is a strong determinant of taxonomic richness—Rook et al., 2013) that preferentially recycled the sediments, destroying the remains of preserved organisms, and the habitats (such as the coastal swamps that supported large and luxuriant growth of terrestrial flora, as evidenced by the large tree trunks and pebble-gravel sized petrified wood that make bulk of the embedded sediments) or periodic subaerial exposure and karstification or evaporite deposition (such as the Kallakkudi Sandstone Member and Karai Formation respectively) or the prevalent high-intensity ravinement events that eroded and redeposited the older sediments (as evident along the contact between Karai and Sillakkudi formations), or the prevalence of higher oxygenating conditions and the absence of syndepositional cementation during siliciclastics or the combinations of all these might have provided relatively lesser fossilization conditions. It is to be noted that despite the prevalence of certain environmental conditions, settings, and events such as enhanced oxygenation and turbidity, tidal regime, and storm events and associated erosion–redeposition during the deposition of carbonates, they (carbonates) provided higher fossilization potential to the biota.

High rate of habitat change was not a constraint as the opportunistic colonizers were quick to exploit the newly available habitats. For example, when the inner and middle shelf regions of Maastrichtian Sea were affected by storm events, the burrowers (*Ophiomorpha irregulaire*) colonized the newly available shallow high-energy regions and were gradually replaced by the *gryphea* with reestablishment of inner shelf conditions. As observed by Stilwell (2003) during examination of molluscan fauna of southern hemisphere across K/Pg, it is a common phenomenon that once a vacant ecospace is made available either due to catastrophic displacement of native taxa or by extinction due to intrinsic or extrinsic causes, the complex processes of biotic recovery and faunal rebound and the infilling of vacant ecospace take place as a continuum.

On a geologic time scale, consistency/stability of habitat was not conducive of diversity and abundance when other restraining factors are in operation; for example, the Sillakkudi Formation and the Karai Formation, ranging thickness up to 400 and 450 m, respectively, and covering relatively many orders of areal extents when compared with other formations, were outpaced by the Kallankurichchi Formation (40 m thick) and the Grey Shale Member (only 7 m thick) in terms of diversity and abundance. Nevertheless, lithological bias and prevalent conducive fossilization milieu, in addition to lifestyle and other environmental factors, might have also played role in these observed disparities. On a basinal scale, the depositional and climatic conditions were highly fluctuating prior to the Santonian, which are reflected in higher habitat heterogeneity and species turnover. Persistent and secular trends were prevalent since Santonian and are reflected in lower habitat heterogeneity, higher species diversity, and population.

7.2 LOCAL–REGIONAL–GLOBAL IMPLICATIONS

The results and conclusions emanating from the case study as well as the review presented in previous chapters, three levels of implications emerged that range from local to global scale. The sequence stratigraphic concepts espoused the control exercised by tectonics–climate-relative sea-level fluctuations-sediment influx/production as the reason for shoreline dynamics, facies distribution, and stacking pattern in spatial and temporal realms. Numerous studies have affirmed these controls and resultant facies distribution patterns and also examined their dynamics on spatial and temporal scales. However, the studies that followed have largely overlooked the implication of these controls on biotic realms, as the shoreline advancement or retreat resulting from eustasy, tectonics, or other causes impact significantly the environmental parameters and spatial and temporal distribution of ecological niches, and thereby affect the occurrence, distribution, diversity, and population of biota. In this book, we emphasized the importance of recognizing this largely relegated field of geosciences. It is demonstrated through a case study that there exists a close coupling between climate–tectonics-relative sea-level fluctuations-sediment influx/production and biotic response as a function of spatiotemporal dynamics of habitats and environmental parameters of ecospaces. As the retreat or advancement of shoreline directly alters the habitat availability and controls the environmental conditions therein (either adverse or conducive) and forces the biota to shift to newer available spaces, or adopt to newer conditions, or dwindle or diversify or to become extinct, studies are required to be conducted on a variety of environmental, paleogeographic, and chronological settings for thorough understanding of this phenomenon.

It is demonstrated through a case study that the factors that control sequence development can be recognized through systematic study of spatial–temporal facies pattern, contact relationship, sedimentary, and tectonic structures and associated features. Enlisting of these traits help construct relative sea-level fluctuations curve, with which, the relative dominance of prevalent geological events, including tectonics, relative sea-level, and others, over facies pattern development could be affirmed. In other words, the habitat dynamics in terms of shift/enlargement/reduction/alteration of depositional systems/ecospaces/habitats are recognizable through spatial–temporal changes of lithofacies and associated biotic changes in terms of shift, dwindling, adaptation, proliferation, and extinction, as the case may be. Only a few studies have been conducted so far (e.g., Ruban et al., 2015) that have documented and lucidly demonstrated the habitat-biodiversity dynamics on a local, regional, and global scale as a function of regional shoreline shifts, basin depth changes, paleobiogeographical changes, and plate tectonic reorganizations.

7.3 FUTURE TRENDS AND RECOMMENDATIONS

Due consideration need to be given in the habitat analysis to the lithological bias that overrides consistency and stability of habitats and environmental conditions. Despite the prevalence of conducive conditions of habitat diversity and sustenance, the lifestyle of organisms and fossilization/preservation conditions also play role. Thus, similar to the studies on biodiversity analysis, habitats need to be studied and documented, including the potential bias that may enhance/destroy the geological patterns of habitat existence/absence, and distribution.

Through perusal of the available literature in the field of research on habitats, it is brought out that the scale of habitat, dynamics, and heterogeneity is used only in a relative context. Hence, establishment of standard reference delimitations and definitions and glossary for habitat are necessary.

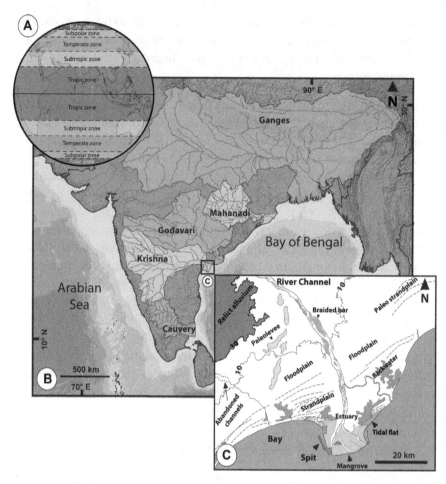

Figure 7.1 **Conceptual depictions of three-fold classification of habitats.** *As depicted in (A), macro habitats can be defined by latitudinal climatic zones. The meso habitat can be delimited by regional landforms such as river basins or deltas (B), while the micro habitats can be the individual subenvironments or landforms, such as flood plain, spit, mangrove swamp, tidal flat, and others as in (C).*

A three-tier classification of habitats in terms of "macro, meso, and micro" habitats is proposed (Fig. 7.1). Accordingly, the latitudinal climatic separations are termed as "macro" habitats, the regional landforms/habitats such as mountains, river basins, and others are termed as "meso" habitats and the field-scale landforms such as tidal flats, flood plain, river channel, estuary, and others are termed as "micro" habitats. This is relative in spatial domain. Nevertheless, this may serve as a starting point of discussion among scientific community and may lead to refinement and definitive explanation and practice.

It is also established by our data and numerous case studies as presented in the previous publications cited in the book that habitat should have recognizable geomorphic boundary/terminology, within which unique physical, chemical, and biological conditions prevail. In geological sense, it is defined as depositional environment and depositional system and others. As each unique depositional environment/depositional system produces unique facies type and their successive occurrences during geological ages produce facies succession/bundle, recognition of facies types and successions, separated by unconformity surfaces or correlative surfaces, as practiced in sequence stratigraphy, provides a convenient measure of habitat heterogeneity and diversity. It is explained succinctly that, the controlling factors that influence sequence development are those factors that influence the habitat as well. Hence, the facies, facies succession and pattern can be considered as proxies to document habitat diversity of geological past. It is practicable at a temporal scale of choice/sampling. Also detailed in the book is the availability of geochemical proxies to document the paleoenvironmental, facies, and habitat dynamics. As succinctly demonstrated by multitudes of chemostratigraphic studies elsewhere, and also due to the nature of many of the chemostratigraphic signals that are global in nature, there exists a possibility of comparison and correlation of habitats so recognized at a variety of spatial scales. With all these, we hypothesize the nexus between tectonics—climate—sediment production/influx-relative sea-level fluctuation that produces unique identifiable facies patterns in geological history, and forms unique habitat type, diversity and pattern, the spatial and temporal scale of which can be tracked through multiproxy studies. Application of this methodology to diverse depositional settings and environments may prove to be a reliable tool for recognition, comparison, and correlation of habitats.

REFERENCES

Acharyya, S.K., Lahiri, T.C., 1991. Cretaceous palaeogeography of the Indian subcontinent: a review. Cretaceous Res. 12, 3−26.

Albano, P.G., 2014. Comparison between death and living land mollusk assemblages in six forested habitats in northern Italy. Palaios 29. Available from: http://dx.doi.org/10.2110/palo.2014.020.

Alvarez, N.O.C., Roser, B.P., 2007. Geochemistry of black shales from the Lower Cretaceous Paja Formation, Eastern Cordillera, Colombia: source weathering, provenance, and tectonic setting. J. South Am. Earth Sci. 23, 271−289.

Anderson, L.D., Delaney, M.L., Faul, K.L., 2001. Carbon to phosphorus ratios in sediments: implications for nutrient cycling. Global Biogeochem. Cycles 15, 65−79.

Ando, A., Huber, B.T., MacLeod, K.G., 2010. Depth-habitat reorganization of planktonic foraminifera across the Albian/Cenomanian boundary. Paleobiology 36. Available from: http://dx.doi.org/10.1666/09027.1.

Andreozzi, M., Dinelli, E., Tateo, F., 1997. Geochemical and mineralogical criteria for the identification of ash layers in the stratigraphic framework of a foredeep: the Early Miocene Mt. Cervarola sandstones, northern Italy. Chem. Geol. 137, 23−39.

Andrew, A.S., Whitford, D.J., Hamilton, P.J., Scarano, S. and Buckley, M., 1996 Application of chemostratigraphy to petroleum exploration and field appraisal. An example from the Surat Basin. Proceedings of Seminar on Asia Pacific Oil and Gas, 421−429.

Archuby, F.M., Adami, M., Martinelli, J.C., Gordillo, S., Boretto, G.M., Malve, M.E., 2015. Regional-scale compositional and size fidelity of rocky intertidal communities from the Patagonian Atlantic coast. Palaios 30. Available from: http://dx.doi.org/10.2110/palo.2014.054.

Ayyasami, K., 2006. Role of oysters in biostratigraphy: a case study from the Cretaceous of the Ariyalur area, southern India. Geosci. J. 10, 237−247.

Ayyasami, K., Das, I., 1990. Unusual preservation of a Cretaceous turtle fossil. J. Geol. Soc. India 36, 519−522.

Ayyasami, K., Rao, B.R.J., 1987. New belemnoid from the Cretaceous of South India. Geol. Surv. India Special Publ. 11, 409−412.

Badgley, C., Finarelli, J.A., 2013. Diversity dynamics of mammals in relation to tectonic and climatic history: comparison of three Neogene records from North America. Paleobiology 39. Available from: http://dx.doi.org/10.1666/12024.

Balseiro, D., Waisfeld, B.G., Vaccari, N.E., 2011. Paleoecological dynamics of Furongian (Late Cambrian) trilobite-dominated communities from northwestern Argentina. Palaios 26. Available from: http://dx.doi.org/10.2110/palo.2010.p10-152r.

Bambach, R.K., 1977. Species richness in marine benthic habitats through the Phanerozoic. Paleobiology 3, 152−167.

Banerji, R.K., 1972. Stratigraphy and micropalaeontology of the Cauvery Basin. Part I, exposed area. J. Palaeontol. Soc. India 17, 1−24.

Baucon, A., Neto de Carvalho, C., 2016. Stars of the aftermath: asteriacites beds from the Lower Triassic of the Carnic Alps (Werfen Formation, Sauris di Sopra), Italy. Palaios 31. Available from: http://dx.doi.org/10.2110/palo.2015.015.

Bellanca, A., Erba, E., Neri, R., Silva, I.P., Sprovieri, M., Tremolada, F., et al., 2002. Palaeoceanographic signifiance of the Tethyan 'Livello Selli' (Early Aptian) from the Hybla Formation, northwestern Sicily: biostratigraphy and high-resolution chemostratigraphic records. Palaeogeogr. Palaeoclimatol. Palaeoecol. 185, 175–196.

Benton, M.J., 2013. Origins of biodiversity. Palaeontology 56, 1–7.

Bhatia, S.B., 1984. Ostracod faunas of the Indian subcontinent-their palaeozoogeographic and palaeoecologic implications. J. Palaeontol. Soc. India 20, 1–8.

Bijma, J., Pörtner, H.-O., Yesson, C., Rogers, A.D., 2013. Climate change and the oceans—what does the future hold? Mar. Pollut. Bull. 74, 495–505.

Blanford, H.F., 1862. On the Cretaceous and other rocks of South Arcot and Trichinopoly districts. Mem. Geol. Surv. India 4, 1–217.

Blum, M., Martin, J., Milliken, K., Garvin, M., 2013. Paleovalley systems: insights from quaternary analogs and experiments. Earth Sci. Rev. 116, 128–169.

Böhm, F., Westphal, H., Bornholdt, S., 2003. Required but disguised: environmental signals in Limestone-marl alternations. Palaeogeogr. Palaeoclimatol. Palaeoecol. 189, 161–178.

Bonis, N.R., Ruhl, M., Kürschner, W.M., 2010. Milankovitch-scale palynological turnover across the Triassic–Jurassic transition at St. Audrie's Bay, SW UK. J. Geol. Soc. London 167. Available from: http://dx.doi.org/10.1144/0016-76492009-141.

Boulila, S., de Rafélis, M., Hinnov, L.A., Gardin, S., Galbrun, B., Collin, P., 2010. Orbitally forced climate and sea-level changes in the Paleoceanic Tethyan domain (marl–limestone alternations, Lower Kimmeridgian, SE France). Palaeogeogr. Palaeoclimatol. Palaeoecol. 292, 57–70.

Bracchi, V., Savini, A., Marchese, F., Palamara, S., Basso, D., Corselli, C., 2015. Coralligenous habitat in the Mediterranean Sea: a geomorphological description from remote data. Ital. J. Geosci. 134. Available from: http://dx.doi.org/10.3301/IJG.2014.16.

Brady, M., 2016. Middle to Upper Devonian skeletal concentrations from carbonate-dominated settings of North America: investigating the effects of bioclast input and burial rates across multiple temporal and spatial scales. Palaios 31. Available from: http://dx.doi.org/10.2110/palo.2015.076.

Bragina, L.G., Bragin, N.Yu, 2013. New data on the Albian–Cenomanian radiolarians from Karai Formation (South India). Stratigr. Geol. Correl. 21, 515–530.

Brayard, A., Nützel, A., Stephen, D.A., Bylund, K.G., Jenks, J., Bucher, H., 2010. Gastropod evidence against the Early Triassic Lilliput effect. Geology 38, 147–150.

Brom, K.R., Salamon, M.A., Ferré, B., Brachaniec, T., Szopa, K., 2015. The Lilliput effect in crinoids at the end of the Oceanic Anoxic Event 2: a case study from Poland. J. Paleontol. 89. Available from: http://dx.doi.org/10.1017/jpa.2016.10.

Cantalapiedra, J.L., Hernandez-Fernandez, M., Alcalde, G., Azanza, B., DeMiguel, D., Morales, J., 2012. Ecological correlates of ghost lineages in ruminants. Paleobiology 38. Available from: http://dx.doi.org/10.1666/09069.1.

Catuneanu, O., 2006. Principles of Sequence Stratigraphy. Elsevier B.V, Amsterdam.

Catuneanu, O., Galloway, W.E., Kendall, C.G., St, C., Miall, A.D., Posamentier, H.W., et al., 2011. Sequence stratigraphy: methodology and nomenclature. Newsl. Stratigr. 44, 173–245.

Chandra, P.K., 1991 Sedimentary basins of India. Presidential address of VIII Convention of Indian Geological Conress, 1–31.

Chandrasekaran, V.A., Ramkumar, M., 1993. On the occurrence of *Planolites* from the Kallankurichchi Formation, Upper Cretaceous (Ariyalur Group), Tiruchy district, South India and its palaeoenvironmental significance. J. Geol. Assoc. Res. Centre 2, 31–36.

Chandrasekaran, V.A., Ramkumar, M., Jacob, M., Saksena, S., 1993. A preliminary note on the occurrence of *Serpula Socialis* from the Sillakkudi Formation (Campanian), Ariyalur Group, South India. J. Geol. Assoc. Res. Centre 2, 32–36.

Chari, M.V.N., Sahu, J.N., Banerjee, B., Zutshi, P.L., Chandra, K., 1995. Evolution of the Cauvery Basin, India from subsidence modeling. Mar. Pet. Geol. 12, 667–675.

Chatterjee, S., Goswami, A., Scotese, C.R., 2013. The longest voyage: tectonic, magmatic, and paleoclimatic evolution of the Indian plate during its northward flight from Gondwana to Asia. Gondwana Res. 23, 238–267.

Chen, Z.-Q., Joachimski, M., Montañez, I., Isbell, J., 2014. Deep time climatic and environmental extremes and ecosystem response: an introduction. Gondwana Res. 25, 1289–1293.

Cherns, L., Wheeley, J.R., Wright, V.P., 2008. Taphonomic windows and molluscan preservation. Palaeogeogr. Palaeoclimatol. Palaeoecol. 270, 220–229.

Chiplonkar, G.W., Tapaswi, P.M., 1979. Biostratigraphy, age and affinities of the bivalve fauna of the Cretaceous of Tiruchirapalli District, South India. Geol. Surv. India Miscellaneous Publ. 45, 137–164.

Clapham, M.E., James, N.P., 2012. Regional-scale marine faunal change in eastern Australia during Permian climate fluctuations and its relationship to local community restructuring. Palaios 27. Available from: http://dx.doi.org/10.2110/palo.2012.p12-003r.

Coccioni, R., Galeotti, S., 2003. The mid-Cenomanian event: prelude to OAE 2. Palaeogeogr. Palaeoclimatol. Palaeoecol. 190, 427–440.

Connell, J.H., 1978. Diversity in tropical rain forests and coral reefs. Science 199, 1302–1310.

Connell, J.H., 1997. Disturbance and recovery of coral assemblages. Coral Reefs 16, S101–S113.

Cook, T.D., Eaton, J.G., Newbrey, M.G., Wilson, M.V.H., 2014. A new genus and species of freshwater stingray (Myliobatiformes, Dasyatoidea) from the latest Middle Eocene of Utah, U.S.A. J. Paleontol. 88. Available from: http://dx.doi.org/10.1666/13-046.

Crampton, J.S., Foote, M., Cooper, R.A., Beu, A.G., Peters, S.E., 2011. The fossil record and spatial structuring of environments and biodiversity in the Cenozoic of New Zealand. Geol. Soc. Spec. Publ. 358, 105–122.

Darroch, S.A.F., 2012. Carbonate facies control on the fidelity of surface-subsurface agreement in benthic foraminiferal assemblages: implications for index-based paleoecology. Palaios 27. Available from: http://dx.doi.org/10.2110/palo.2011.p11-027r.

Das, N., 1997. Chemostratigraphy of sedimentary sequences: a review of the state of the art. J. Geol. Soc. India 49, 621–628.

Diaz-Martinez, I., Suarez-Hernando, O., Martinez-Garcia, B., Hernandez, J.M., Fernandez, S.G., Perez-Lorente, F., et al., 2015. Early Miocene shorebird-like footprints from the Ebro Basin, La Rioja, Spain: paleoecological and paleoenvironmental significance. Palaios 30. Available from: http://dx.doi.org/10.2110/palo.2014.078.

Domingo, M.S., Badgley, C., Azanza, B., DeMiguel, D., Alberdi, M.T., 2014. Diversification of mammals from the Miocene of Spain. Paleobiology 40. Available from: http://dx.doi.org/10.1666/13043.

Dubicka, Z., Peryt, D., Szuszkiewicz, M., 2014. Foraminiferal evidence for paleogeographic and paleoenvironmental changes across the Coniacian-Santonian boundary in western Ukrine. Palaeogeogr. Palaeoclimatol. Palaeoecol. 401, 43–56.

Egerton, P., de, M.G., 1845. On the remains of fishes found by Mr. Kaye and Mr. Cunliffe in the Pondicherry Beds. J. Geol. Soc. London 1, 164–171.

Faith, J.T., Behrensmeyer, A.K., 2013. Climate change and faunal turnover: testing the mechanics of the turnover-pulse hypothesis with South African fossil data. Paleobiology 39. Available from: http://dx.doi.org/10.1666/12043.

Fenner, J., 2001a. Palaeoceanographic and climatic changes during the Albian: summary of the results from the Kirchrode boreholes. Palaeogeogr. Palaeoclimatol. Palaeoecol. 174, 287–304.

Fenner, J., 2001b. The Kirchrode I and II boreholes: technical details and evidence on tectonics and on palaeoceanographic development during Albian. Palaeogeogr. Palaeoclimatol. Palaeoecol. 174, 33–65.

Flessa, K., Cutler, A.H., Meldahl, K.H., 1993. Time and taphonomy: quantitative estimates of time-averaging and stratigraphic disorder in a shallow marine habitat. Paleobiology 19, 266–286.

Fraiser, M.L., Bottjer, D.J., 2004. The nonactualistic Early Triassic gastropod fauna: a case study of the Lower Triassic Sinbad Limestone member. Palaios 19, 259–275.

Friedman, G.M., Chakraborty, C., 1997. Stable isotopes in marine carbonates: their implications for the paleoenvironment with special reference to the Proterozoic Vindhyan carbonates (Central India). J. Geol. Soc. India 50, 131–159.

Gale, A.S., Smith, A.B., Monks, N.E.A., Young, J.A., Howard, A., Wray, D.S., et al., 2000. Marine biodiversity through the Late Cenomanian—Early Turonian: palaeoceanographic controls and sequence stratigraphic biases. J. Geol. Soc. 157, 745–757.

Garcia, T., Velo, A., Fernandez-Bastero, S., Gago-Duport, L., Santos, A., Alejo, I., et al., 2004. Coupled transport-reaction pathways and distribution patterns between siliciclastic-carbonate sediments at the Ria de Vigo. J. Mar. Syst. Available from: http://dx.doi.org/10.1016/j.jmarsys.2004.07.014.

Govindan, A., Ananthanarayanan, S., Vijayalakshmi, K.G., 2000. Cretaceous petroleum system in Cauvery Basin, India. In: Govindhan, A. (Ed.), Cretaceous Stratigraphy—An Update. Memoirs of Geoloical Society of India, Bangaluru, 46. pp. 365–382.

Govindan, A., Ravindran, C.N., 1996. Cretaceous biostratigraphy and sedimentation history of Cauvery Basin, India. In: Pandey, J., Azmi, R.J., Bhandari, A., Dave, A. (Eds.), Contribution of XV Indian Colloquium on Micropalaeontology and Stratigraphy, Dehra Dun, India, 19–31.

Govindan, A., Ravindran, C.N., Rangaraju, M.K., 1996. Cretaceous stratigraphy and planktonic foraminiferal zonation in Cauvery Basin, South India. Mem. Geol. Soc. India 37, 155–187.

Gowda, S.S., 1964. Fossil fish ossiculiths from the Cenomanian of South India. Eclogae Geol. Helv. 57, 743–746.

Gowda, S.S., 1966. The first fossil otolith from India. Bull. Geol. Soc. India 4, 15–17.

Gowda, S.S., 1967. On a new fossil fish known from an otolith from the South Indian Cenomanian. J. Geol. Soc. India 8, 119–129.

Gradstein, F.M., Ogg, J.G., Smith, A.G. (Eds.), 2004. A New Geologic Time Scale 2004. Cambridge University Press, Cambridge, 464 pp.

Grammer, G.M., Eberli, G.P., Van Buchem, F.S.P., 1996. Application of high resolution sequence stratigraphy to evaluate lateral variability in outcrop and subsurface—Desert Creek and Ismay intervals, Paradox basin. In: Longman, M.W., Sonnelfeld, M.D. (Eds.), Paleozoic Systems of the Rocky Mountain Region, Rocky Mountain Section. Society of Economic Palaeontologists and Mineralogists, USA, pp. 235–266.

Guha, A.K., 1987. Palaeoecology of some Upper Creaceous sediments of India—an approach based on bryozoa. Geol. Surv. India Spec. Publ. 11, 419–429.

Guha, A.K., Senthilnathan, D., 1990. Onychocellids (Bryozoa: Cheilostomata) from the Ariyalur carbonate sediments of South India. J. Palaeontol. Soc. India 35, 41–51.

Guha, A.K., Senthilnathan, D., 1996. Bryozoan fauna of the Ariyalur Group (Late Cretaceous) Tamil Nadu and Pondicherry, India. Palaeontol. Indica 49, 2–17.

Hallam, A., Wignall, P.B., 1999. Mass extinctions and sea-level changes. Earth Sci. Rev. 48, 217–250.

Hancock, J., 2001. A proposal for a new position for the Aptian/Albian boundary. Cretaceous Res. 22, 677–683.

Haq, B.U., 2014. Cretaceous eustasy revisited. Glob. Planet. Change 113, 44–58.

Haq, B.U., Hardenbol, J., Vail, P.R., 1987. Chronology of fluctuating sea levels since the Triassic. Science 235, 1156–1167.

Haq, B.U., Hardenbol, J., Vail, P.R., 1988. Mesozoic and Cenozoic chronostratigraphy and cycles of sea-level change. In: Wilgus, C.K., Hastings, B.S., Posamentier, H., Van Wagoner, J., Ross, C.A., and Kendall, C.G.St.C., (Eds.), Sea-Level Changes, Society of Economic Palaeontologists and Mineralogists, 42, 71–108.

Hart, M.B., Tewari, A., Watkinson, M.P., 1996. Wood boring bivalves from the Trichinopoly Sandstone of the Cauvery Basin, south-east India. In: Pandey, J., Azmi, R.J., Bhandari, A., Dave, A., (Eds.), Contributions to the XV Indian Colloquium on Micropalaeontology and Stratigraphy, Dehradun, 529–539.

Hautmann, M., 2014. Diversification and diversity partitioning. Paleobiology 40. Available from: http://dx.doi.org/10.1666/13041.

Hild, E., Brumsack, H.J., 1998. Major and minor element geochemistry of Lower Aptian sediments from the NW German Basin (core Hoheneggelsen KB 40). Cretaceous Res. 19, 615–633.

Hofmann, P., Ricken, W., Schwark, L., Leythaeuser, D., 2001. Geochemical signature and related climatic-oceanographic processes for Early Albian black shales: site 417D, North Atlantic Ocean. Cretaceous Res. 22, 243–257.

Hoffmann, R., Wochnik, A.S., Heinzl, C., Betzler, S.B., Matich, S., Griesshaber, E., et al., 2014. Nanoprobe crystallographic orientation studies of isolated shield elements of the coccolithophore species Emiliania huxleyi. Eur. J. Mineral. 26. Available from: http://dx.doi.org/10.1127/0935-1221/2014/0026-2365.

Holland, S.M., 2012. Sea-level change and the area of shallow-marine habitat: implications for marine biodiversity. Paleobiology 38, 205–217.

Holland, S.M., Christie, M., 2013. Changes in area of shallow siliciclastic marine habitat in response to sediment deposition: implications for onshore-offshore paleobiologic patterns. Paleobiology 39. Available from: http://dx.doi.org/10.1666/12053.

Holland, S.M., Zaffos, A., 2011. Niche conservatism along an onshore-offshore gradient. Paleobiology 37. Available from: http://dx.doi.org/10.1666/10032.1.

Hottinger, L., 1983. Processes determining the distribution of larger foraminifera in space and time. Utrecht Micropaleontol. Bull. 30, 239–254.

< https://en.wikipedia.org/wiki/Habitat>.

Jafar, S.A., Rai, J., 1989. Discovery of Albian nannoflora from type Dalmiapuram Formation, Cauvery Basin, India-Palaeooceanographic remarks. Curr. Sci. 58, 358–363.

Jafer, S.A., 1996. The evolution of marine Cretaceous basins of India: calibration with nannofossil zones. In: Sahni, A. (Ed.), Cretaceous Stratigraphy and Palaeoenvironments. Memoirs of Geoloical Society of India, Bangaluru, 37. pp. 121–134.

Jarzen, D.M., Klug, C., 2010. A preliminary investigation of a Lower to Middle Eocene palynoflora from Pine Island, Florida, USA. Palynology 34. Available from: http://dx.doi.org/10.1080/01916121003737421.

Jeans, C.V., Long, D., Hall, M.A., Bland, D.J., Conford, C., 1991. The geochemistry of the Plenus marls at Dover, England: evidence of fluctuating oceanographic conditions and of glacial control during the development of the Cenomanian-Turonian $\delta^{13}C$ anomaly. Geol. Mag. 128, 603–632.

Jendrzejewski, L., Littke, R., Rullkoetter, K., 2001. Organic geochemistry and depositional history of Upper Albian sediments from the Kirchrode I borehole, northern Germany. Palaeogeogr. Palaeoclimatol. Palaeoecol. 174, 107–120.

Johnson, K.G., Hasibuan, F., Müller, W., Todd, J.A., 2015. Biotic and environmental origins of the Southeast Asian marine biodiversity hotspot: The throughflow project. Palaios 30. Available from: http://dx.doi.org/10.2110/palo.2014.103.

Kale, A.S., 2011. Comments on 'Sequence surfaces and paleobathymetric trends in Albian to Maastrichtian sediments of Ariyalur area, Cauvery Basin, India' from Nagendra, Kannan, Sen, Gilbert, Bakkiaraj, Reddy, and Jaiprakash. Mar. Pet. Geol. 28, 1252–1259.

Kale, A.S., Lotfalikani, A., Phansalkar, V.G., 2000. Calcareous nanofossils from the Uttatur Group of Trichinopoly Cretaceous, South India. In: Govindan, A. (Ed.), Cretaceous Stratigraphy—An Update. Memoirs of Geological Society of India, Bangaluru, 46. pp. 213–227.

Kale, A.S., Phansalkar, V.G., 1992a. Calcareous nannofossils from the Uttatur Group, Trichinopoly District, Tamil Nadu, India. J. Palaeontol. Soc. India 37, 85–102.

Kale, A.S., Phansalkar, V.G., 1992b. Nannofossil biostratigraphy of the Uttatur Group, Trichinopoly District, South India. Mem. Geol. Soc. India 43, 89–107.

Kalmar, A., Currie, D.J., 2010. The completeness of the continental fossil record and its impact on patterns of diversification. Paleobiology 36. Available from: http://dx.doi.org/10.1666/0094-8373-36.1.51.

Kampschulte, A., Bruckschen, P., Strauss, H., 2001. The sulphur isotope composition of trace sulphates in Carboniferous brachiopods: implications for coeval seawater correlation with other geochemical cycles and isotope stratigraphy. Chem. Geol. 175, 149–173.

Keller, G., Berner, Z., Adatte, T., Stüben, D., 2004. Cenomanian–Turonian and δ^{13}C, and δ^{18}O, sea level and salinity variations at Pueblo, Colorado. Palaeogeogr. Palaeoclimatol. Palaeoecol. 211, 19–43.

Kennedy, M.J., Droser, M.L., 2011. Early Cambrian metazoans in fluvial environments, evidence of the non-marine Cambrian radiation. Geology 39. Available from: http://dx.doi.org/10.1130/G32002.1.

Klug, C., Döring, S., Korn, D., Ebbighausen, 2006. The Viséan sedimentary succession at the Gara el Itima (Anti-Atlas, Morocco) and its ammonoid faunas. Fossil Rec. 9, 3–60.

Kohring, R., Bandel, K., Kortum, D., Parthasarathy, S., 1996. Shell structure of a dinosaur egg from the Maastrichtian of Ariyalur (southern India). Neues Jahrbuch Geologisch Paläontolosch Monat 1, 48–64.

Kossmat, F., 1895. On the importance of the Cretaceous rocks of southern India in estimating the geological conditions during the later Cretaceous times. Rec. Geol. Surv. India 28, 39–55.

Kröger, B., Ebbestad, J.O.R., Lehnert, O., 2016. Accretionary mechanisms and temporal sequence of formation of the Boda Limestone Mud-Mounds (Upper Ordovician), Siljan District, Sweden. J. Sediment. Res. 86. Available from: http://dx.doi.org/10.2110/jsr.2016.12.

Kumar, S.P., 1983. Geology and hydrocarbon prospects of Krishna, Godavari and Cauvery basins. Petrol. Asia J. 6, 57–65.

Lal, N.K., Siawal, A., Kaul, A.K., 2009. Evolution of east coast of India—a plate tectonic reconstruction. J. Geol. Soc. India 73, 249–260.

Leonard-Pingel, J.S., Jackson, J.B.C., 2016. Drilling predation increased in response to changing environments in the Caribbean Neogene. Paleobiology 42. Available from: http://dx.doi.org/10.1017/pab.2016.2.

Littke, R., Jendrzejewski, L., Lokay, P., Shuangqing, W., Rullkoetter, J., 1998. Organic geochemistry and depositional history of the Barremian—Aptian boundary interval in the Lower Saxony Basin, northern Germany. Cretaceous Res. 19, 581–614.

MacArthur, R.H., MacArthur, J.W., 1961. On bird species diversity. Ecology 42, 594–598.

MacLeod, K.G., 1994. Bioturbation, inoceramid extinction, and mid-Maastrichtian ecological change. Geology 22, 139–142.

Mamgain, V.D., Sastry, M.V.A., Subbaraman, J.V., 1973. Report of ammonites from Gondwana plant beds at Terani, Tiruchirapalli District, Tamil Nadu. J. Geol. Soc. India 14, 189–200.

Mancosu, A., Nebelsick, J.H., 2015. The origin and paleoecology of clypeasteroid assemblages from different shelf settings of the Miocene of Sardinia, Italy. Palaios 30. Available from: http://dx.doi.org/10.2110/palo.2014.087.

Martin, R.E., Quigg, A., Podkovyrov, V., 2008. Marine biodiversification in response to evolving phytoplankton stoichiometry. Palaeogeogr. Palaeoclimatol. Palaeoecol. 258, 277–291.

Matley, C.A., 1929. The Cretaceous dinosaurs of the Trichinopoly District and the rocks associated with them. Rec. Geol. Surv. India 61, 337–349.

McClure, K.J., Lockwood, R., 2015. Relationships among *Venericardia* (Bivalvia: Carditidae) on the U.S. Coastal Plain during the Paleogene. J. Paleontol. 89. Available from: http://dx.doi.org/10.1017/jpa.2015.23.

McKirdy, D.M., Burgess, J.M., Lemon, N.M., Yu, X., Cooper, A.M., Gostin, V.A., et al., 2001. A chemostratigraphic overview of the late Cryogenian interglacial sequence in the Adelaide fold thrust belt, South Australia. Precambrian. Res. 106, 149–186.

McMullen, S.K., Holland, S.M., O'Keefe, F., 2014. The occurrence of vertebrate and invertebrate fossils in a sequence stratigraphic context: the Jurassic Sundance Formation, Bighorn Basin, Wyoming, U.S.A. Palaios 29. Available from: http://dx.doi.org/10.2110/pal.2013.132.

Menier, D., Augris, C., Briend, C., 2014a. Les réseaux fluviatiles anciens du plateau continental de Bretagne Sud. 101 pages (Editions QUAE, 2014).

Menier, D., Pierson, B., Chalabi, A., King King, T., Pubellier, M., 2014b. Morphological indicators of structural control, relative sea-level fluctuations and platform drowning on Present-Day and Miocene carbonate platforms. Mar. Pet. Geol. 58, 776–788. Available from: http://dx.doi.org/10.1016/j.marpetgeo.2014.01.016.

Menier, D., Estournès, G., Mathew, M., Ramkumar, M., Briend, C., Siddiqui, N., et al., 2016. Relict geomorphological and structural control on the coastal sediment partitioning, north of Bay of Biscay. Zeitschrift für Geomorphologie 60, 67–74. Available from: http://dx.doi.org/10.1127/zfg/2016/0267.

Mihaljević, M., Renema, W., Welsh, K., Pandolfi, J.M., 2014. Eocene–Miocene shallow-water carbonate platforms and increased habitat diversity in Sarawak, Malaysia. Palaios 29, 378–391.

Misra, P.K., Kishore, S., Singh, S.K., Jauhri, A.K., 2009. Rhodophycean Algae from the Lower Cretaceous of the Cauvery Basin, South India. J. Geol. Soc. India 73, 325–334.

Montagne, A., Naim, O., Tourrand, C., Pierson, B., Menier, D., 2013. Status of coral reef communities on two carbonates platforms (Tun Sakaran Marine Park, east Sabah, Malaysia). J. Ecosyst. Available from: http://dx.doi.org/10.1155/2013/358183.

Montero-Serrano, J.C., Palarea-Albaladejo, J., Martín-Fernández, J.A., Martínez-Santana, M., Gutiérrez-Martín, J.V., 2010. Sedimentary chemofacies characterization by means of multivariate analysis. Sediment. Geol. 228, 218–228.

Munnecke, A., Westphal, H., 2004. Shallow water aragonite recorded in bundles of limestone-marl alternations—the Upper Jurassic of SW Germany. Sediment. Geol. 164, 191–202.

Myers, C.E., Stigall, A.L., Lieberman, B.S., 2015. PaleoENM: applying ecological niche modeling to the fossil record. Paleobiology 41. Available from: http://dx.doi.org/10.1017/pab.2014.19.

Narayanan, V., 1977. Biozonation of the Utatur Group, Trichinopoly, Cauvery Basin. J. Geol. Soc. India 18, 415–428.

Neto, J.V.Q., Sames, B., Colin, J.P., 2014. *Kegelina*: a new limnic ostracod (Cyprideidae, Cypridoidea) genus from the Lower Cretaceous of the Americas and Africa. J. Paleontol. 88. Available from: http://dx.doi.org/10.1666/13-019.

Novack-Gottshall, P.M., 2016. General models of ecological diversification. I. Conceptual synthesis. Paleobiology 42. Available from: http://dx.doi.org/10.1017/pab.2016.3.

Novak, V., Renema, W., 2015. Larger foraminifera as environmental discriminators in Miocene mixed carbonate–siliciclastic systems. Palaios 30. Available from: http://dx.doi.org/10.2110/palo.2013.081.

Nürnberg, S., Aberhan, M., 2013. Habitat breadth and geographic range predict diversity dynamics in marine Mesozoic bivalves. Paleobiology 39. Available from: http://dx.doi.org/10.1666/12047.

O'Dogherty, L., Sandoval, J., Vera, J.A., 2000. Ammonite faunal turnover tracing sea-level changes during the Jurassic (Betic Cordillera, southern Spain). J. Geol. Soc. 157, 723–736.

Paranjape, A.R., Kulkarni, K.G., Kale, A.S., 2014. Sea-level changes in the Upper Aptian-Lower/Middle(?) Turonian sequence of Cauvery Basin, India—an ichnological perspective. Cretaceous Res.1–14. Available from: http://dx.doi.org/10.1016/j.cretres.2014.11.005.

Paul, S.N., 1973. A note on the fossil shark tooth from Tiruchchirappalli District, Tamil Nadu, India. Curr. Sci. 42, 753.

Payne, J.L., 2005. Evolutionary dynamics of gastropod size across the end-Permian extinction and through the Triassic recovery interval. Paleobiology 31, 269–290.

Pearce, M.A., Jarvis, I., Swan, A.R.H., Murphy, A.M., Tocher, B.A., Edmunds, W.M., 2003. Integrating palynological and geochemical data in a new approach to palaeoecological studies: Upper Cretaceous of the Banterwick Barn Chalk borehole, Berkshire, UK. Mar. Micropaleontol. 47, 271–306.

Pellenard, P., Tramoy, R., Puceat, E., Huret, E., Martinez, M., Bruneau, L., et al., 2014. Carbon cycle and sea-water palaeotemperature evolution at the Middle-Late Jurassic transition, eastern Paris Basin (France). Mar. Pet. Geol. 53, 30–43.

Perry, C.T., Kench, P.S., O'Leary, M.J., Morgan, K.M., Januchowski-Hartley, F., 2015. Linking reef ecology to island building: parrot fish identified as major producers of island-building sediment in the Maldives. Geology 43, 503–506.

Peryt, D., Wyrwicka, K., 1993. The Cenomanian/Turonian boundary event in Central Poland. Palaeogeogr. Palaeoclimatol. Palaeoecol. 104, 185–197.

Pian, S., Menier, D., Sedrati, M., 2014. Analysis of morphodynamic beach states along the South Brittany coast. Geomorphol. Relief Processus Environ. 3, 261–274.

Powell, C. Mc. A., Roots, S.R., Veevers, J.J., 1988. Pre-break up continental extension in east Gondwanaland and the early opening of the Indian Ocean. Tectonophysics 155, 261–283.

Powers, C.M., Bottjer, D.J., 2009. The effects of mid-Phanerozoic environmental stress on bryozoan diversity, paleoecology, and paleogeography. Glob. Planet. Change 65, 146–154.

Prabhakar, K.N., Zutshi, P.L., 1993. Evolution of southern part of Indian east coast basins. J. Geol. Soc. India 41, 215–230.

Prasad, G.V.R., Verma, O., Flynn, J.J., Goswami, A., 2013. A new Late Cretaceous vertebrate fauna from the Cauvery Basin, South India: implications for Gondwanan paleobiogeography. J. Vertebr. Paleontol. 33, 1260–1268.

Pratt, L.M., 1984. Influence of paleoenvironmental factors on preservation of organic matter in Middle Cretaceous Greenhorn Formation, Puebio, Colorado. Am. Assoc. Petol. Geol. Bull. 68, 1146–1159.

Purdy, E.G., 2008. Comparison of taxonomic diversity, strontium isotope and sea-level patterns. Int. J. Earth Sci. 97, 651–664.

Rachold, V., Brumsack, H.J., 2001. Inorganic geochemistry of Albian sediments from the Lower Saxony Basin, NW Germany: palaeoenvironmental constraints and orbital cycles. Palaeogeogr. Palaeoclimatol. Palaeoecol. 174, 121–143.

Radulović, B.V., Wagih, A., Radulović, V.J., Banjac, N.J., 2015. *Sillakkudihynchia gen. nov.* (Rhynchonellida: Brachiopoda), from the Upper Cretaceous (Campanian), of the Cauvery Basin, southern India: taxonomy, palaeoecology and palaeobiogeography. Neues Jahrbuch Geologisch Paläontologie 276, 63–78.

Rai, R., Ramkumar, M., Sugantha, T., 2013. Calcareous nannofossils from the Ottakoil Formation, Cauvery Basin, South India: implications on age, biostratigraphic correlation and palaeobiogeography. In: Ramkumar, M. (Ed.), On the Sustenance of Earth's Resources. Springer-Verlag, Heidelberg, pp. 109–122.

Raju, D.S.N., Ravindran, C.N., 1990. Cretaceous sea-level changes and transgressive/regressive phases in India—a review. Proceedings on Cretaceous event stratigraphy and the correlation of Indian non-marine strata, Contributions to Seminar cum Workshop on IGCP-216, Chandigarh, 38–46.

Raju, D.S.N., Ravindran, C.N., Kalyansundar, R., 1993. Cretaceous cycles of sea-level changes in Cauvery Basin, India—a first revision. Oil Nat. Gas Corp. Bull. 30, 101–113.

Ramanathan, S., 1968. Stratigraphy of the Cauvery Basin with reference to its oil prospects, Cretaceous-Tertiary of South India. Memoirs of Geological Society of India, Bangaluru, 2. pp. 153–167.

Rama Rao, L., 1932. On a reptilian vertebra from the South Indian Cretaceous. Am. J. Sci. 24, 221–224.

Ramkumar, M., 1996. Evolution of Cauvery Basin and tectonic stabilization of parts of South Indian shield—insights from structural and sedimentologic data. J. Geol. Assoc. Res. Centre Misc.Publ. 4, 1–15.

Ramkumar, M., 1997a. Plea for national geological field museum. Indian Assoc. Sedientol. Newsl. 1, 2–3.

Ramkumar, M., 1997b. Ecologic adaptation of *Serpula Socialis*—a study from South Indian Cretaceous sequence. J. Geol. Assoc. Res. Centre 5, 153–158.

Ramkumar, M., 1999. Role of chemostratigraphic technique in reservoir characterization and global stratigraphic correlation. Indian J. Geochem. 14, 33–45.

Ramkumar, M., 2000. Recent changes in the Kakinada spit, Godavari delta. J. Geol. Soc. India 55, 183–188.

Ramkumar, M., 2001. Sedimentary environments of the modern Godavari delta: characterization and statistical discrimination towards computer assisted environment recognition scheme. J. Geol. Soc. India 57, 49–63.

Ramkumar, M., 2003. Progradation of the Godavari delta: a fact or empirical artifice? Insights from coastal landforms. J. Geol. Soc. India 62, 290–304.

Ramkumar, M., 2008. Cyclic fine-grained deposits with polymict boulders in Olaipadi Member of the Dalmiapuram Formation, Cauvery Basin, South India: plausible causes and sedimentation model. J. Earth Sci. 2, 7–27.

Ramkumar, M., 2015a. Marine Paleobiodiversity: Responses to Sea-Level Cycles and Perturbations. Elsevier, Amsterdam, The Netherlands, 54 pp.

Ramkumar, M., 2015b. Cretaceous Sea Level Rise: Down Memory Lane and the Road Ahead. Elsevier, Amsterdam, The Netherlands, 58 pp.

Ramkumar, M., 2015c. Discrimination of tectonic dynamism, quiescence, and third-order relative sea-level cycles of the Cauvery Basin, South India. Ann. Geol. Balkan Peninsula 76, 19–45.

Ramkumar, M., 2015d. Toward standardization of terminologies and recognition of chemostrati-graphy as a formal stratigraphic method. In: Ramkumar, M. (Ed.), Chemostratigraphy: Concepts, Techniques and Applications. Elsevier, pp. 1–21. Available from: http://dx.doi.org/10.1016/B978-0-12-419968-2.00001-7

Ramkumar, M., Berner, Z., 2015. Temporal trends of geochemistry, relative sea-level and source area weathering in the Cauvery Basin, South India. In: Ramkumar, M. (Ed.), Chemostratigraphy: Concepts, Techniques and Applications. Elsevier, Amsterdam, The Netherlands, pp. 273–308.

Ramkumar, M., Chandrasekaran, V.A., 1996. Megafauna and environmental conditions of Kallankurichchi Formation (Lower Maestrichtian), Ariyalur Group, Tiruchy district, South India. J. Geol. Assoc. Res. Centre 4, 38–45.

Ramkumar, M., Sathish, G., 2009. Palaeoenvironmental and sequence stratigraphic significance of the occurrence of Ophiomorpha irregulaire in the Kallankurichchi Formation, Ariyalur Group, Cauvery Basin, South India. Palaeontol. Stratigr. Facies 17, 129–137.

Ramkumar, M., Pattabhi Ramayya, M., Gandhi, M.S., 2000a. Beach rock exposures at wave cut terraces of modern Godavari delta: their genesis, diagenesis and indications on coastal submergence and sea-level rise. Indian J. Mar. Sci. 29, 219–223.

Ramkumar, M., Sudha Rani, P., Gandhi, M.S., Pattabhi Ramayya, M., Rajani Kumari, V., Bhagavan, K.V.S., et al., 2000b. Textural characteristics and depositional sedimentary environments of the modern Godavari delta. J. Geol. Soc. India 56, 471–487.

Ramkumar, M., Stüben, D., Berner, Z., 2004a. Lithostratigraphy, depositional history and sea-level changes of the Cauvery Basin, southern India. Ann. Geol. Balkan Peninsula 65, 1–27.

Ramkumar, M., Stüben, D., Berner, Z., Schneider, J., 2004b. Geochemical and isotopic anomalies preceding K/T boundary in the Cauvery Basin, South India: implications for the end Cretaceous events. Curr. Sci. 87, 1738–1747.

Ramkumar, M., Subramanian, V., Stüben, D., 2005a. Deltaic sedimentation during Cretaceous Period in the Northern Cauvery Basin, South India: facies architecture, depositional history and sequence stratigraphy. J. Geol. Soc. India 66, 81–94.

Ramkumar, M., Harting, M., Stüben, D., 2005b. Barium anomaly preceding K/T boundary: plausible causes and implications on end Cretaceous events of K/T sections in Cauvery Basin (India), Israel, NE-Mexico and Guatemala. Int. J. Earth Sci. 94, 475–489.

Ramkumar, M., Stüben, D., Berner, Z., 2006. Elemental interrelationships and depositional controls of Barremian-Danian strata of the Cauvery Basin, South India: implications on scales of chemostratigraphic modelling. Indian J. Geochem. 21, 341–367.

Ramkumar, M., Anbarasu, K., Sugantha, T., Rai, J., Sathish, G., Suresh, R., 2010a. Occurrences of KTB exposures and dinosaur nesting site near Sendurai, India: an initial report. Int. J. Phys. Sci. 22, 573–584.

Ramkumar, M., Berner, Z., Stüben, D., 2010b. Hierarchical delineation and multivariate statistical discrimination of chemozones of the Cauvery Basin, South India: implications on spatio-temporal scales of stratigraphic correlation. Petrol. Sci. 7, 435–447.

Ramkumar, M., Stüben, D., Berner, Z., Schneider, J., 2010c. $^{87}Sr/^{86}Sr$ anomalies in Late Cretaceous-Early Tertiary strata of the Cauvery Basin, South India: constraints on nature and rate of environmental changes across K-T boundary. J. Earth Syst. Sci. 119, 1–17.

Ramkumar, M., Stueben, D., Berner, Z., 2011. Barremian–Danian chemostratigraphic sequences of the Cauvery Basin, India: implications on scales of stratigraphic correlation. Gondwana Res. 19, 291–309.

Ramkumar, M., Sugantha, T., Rai, J., 2013. Facies and granulometric characteristics of the Kallamedu Formation, Ariyalur Group, South India: implications on Cretaceous-Tertiary Boundary events. In: Ramkumar, M. (Ed.), On a Sustainable Future of the Earth's Natural Resources. Springer-Verlag, Heidelberg, pp. 263–284. Available from: http://dx.doi.org/10.1007/978-3-642-32917-3_15

Ramkumar, M., Menier, D., Manoj, M.J., Santosh, M., 2016a. Geological, geophysical and inherited tectonic imprints on the climate and contrasting coastal geomorphology of the Indian peninsula. Gondwana Res. 36, 52–80.

Ramkumar, M., Menier, D., Manoj, M.J., Santosh, M., Siddiqui, N.A., 2016b. Early Cenozoic rapid flight enigma of the Indian subcontinent resolved: Roles of topographic top loading and subcrustal erosion. Geosci. Front. Available from: http://dx.doi.org/10.1016/j.gsf.2016.05.004.

Raymo, E., Oppo, D.W., Curry, W., 1997. The mid-Pleistocene climate transition: a deepsea carbon isotopic perspective. Paleoceanography 12, 546−559.

Reich, S., 2014. Gastropod associations as a proxy for seagrass vegetation in a tropical, carbonate setting (San Salvador, Bahamas). Palaios 29. Available from: http://dx.doi.org/10.2110/palo.2013.071.

Reich, S., Warter, V., Wesselingh, F.P., Zwaan, H., Renema, W., Lourens, L., 2015. Paleoecological significance of stable isotope ratios in Miocene tropical shallow-marine habitats (Indonesia). Palaios 30, 53−65.

Reid, W.V., 1998. Biodiversity hotspots. Trends Ecol. Evol. 13, 275−280.

Reineck, H.E., Singh, I.B., 1980. Depositional sedimentary environments. Springer-Verlag, Heidelberg, 439 pp.

Renema, W., Troelstra, S.R., 2001. Larger foraminifera distribution on a mesotrophic carbonate shelf in SW Sulawesi (Indonesia). Palaeogeogr. Palaeoclimatol. Palaeoecol. 175, 125−146.

Retallack, G.J., 2011. Exceptional fossil preservation during CO_2 greenhouse crises? Palaeogeogr. Palaeoclimatol. Palaeoecol. 307, 59−74.

Retallack, G.J., 2015. How well do fossil assemblages of the Ediacara Biota tell time? Comment. Geology 42. Available from: http://dx.doi.org/10.1130/G34781C.1.

Reymond, C.E., Bode, M., Renema, W., Pandolfi, J., 2011. Ecological incumbency impedes stochastic community assembly in Holocene foraminifera from the Huon Peninsula, Papua New Guinea. Paleobiology 37. Available from: http://dx.doi.org/10.1666/09087.1.

Richey, J.N., Sachs, J.P., 2016. Precipitation changes in the western tropical Pacific over the past millennium. Geology 44. Available from: http://dx.doi.org/10.1130/G37822.1.

Ricketts, T.H., Daily, G.C., Ehrlich, P.R., Fay, J.P., 2015. Countryside biogeography of Moths in a fragmented landscape: biodiversity in native and agricultural habitats. Conserv. Biol. 15, 378−388.

Riegel, W., Lenz, O.K., Wilde, V., 2015. From open estuary to meandering river in a greenhouse world: an ecological case study from the middle Eocene of Helmstedt, Northern Germany. Palaios 30. Available from: http://dx.doi.org/10.2110/palo.2014.005.

Ritterbush, K.A., Bottjer, D.J., Corsetti, F.A., Rosas, S., 2014. New evidence on the role of siliceous sponges in ecology and sedimentary facies development in eastern Panthalassa following theTriassic−Jurassic mass extinction. Palaios 29. Available from: http://dx.doi.org/10.2110/palo.2013.121.

Ritterbush, K.A., Ibarra, Y., Tackett, L.L., 2016. Post-extinction biofacies of the first carbonate ramp of the Early Jurassic (Sinemurian) in NE Panthalassa (New York Canyon, Nevada, USA). Palaios 31. Available from: http://dx.doi.org/10.2110/palo.2015.021.

Roberts, C.M., Ormond, R.F.G., 1987. Habitat complexity and coral reef fish diversity and abundance on Red Sea fringing reefs. Mar. Ecol. Program 41, 1−8.

Romano, M., Whyte, M.A., 2013. A new record of the trace fossil *Selenichnites* from the Middle Jurassic Scalby Formation of the Cleveland Basin, Yorkshire. Proc. Yorkshire Geol. Soc. 59. Available from: http://dx.doi.org/10.1144/pygs2013-329.

Rook, D.L., Heim, N.A., Marcot, J., 2013. Contrasting patterns and connections of rock and biotic diversity in the marine and non-marine fossil records of North America. Palaeogeogr. Palaeoclimatol. Palaeoecol. 372, 123−129.

Rosen, B.R., 1984. Reef Coral Biogeography and Climate through the Late Cainozoic: Just Islands in the Sun or a Critical Pattern of Islands. In: Brenchley, P.J. (Ed.), Fossils and Climate. 201−262.

Roy, K., Jablonski, D., Valentine, J.W., Rosenberg, G., 1998. Marine latitudinal diversity gradients: tests of causal hypotheses. Proc. Natl. Acad. Sci. USA 95, 3699–3702.

Ruban, D.A., 2009. Phanerozoic changes in the high-rank suprageneric diversity structure of brachiopods: linear and non-linear effects. Palaeoworld 18, 263–277.

Ruban, D.A., 2010a. Do new reconstructions clarify the relationships between the Phanerozoic diversity dynamics of marine invertebrates and long-term eustatic trends? Ann. Paléontol. 96, 51–59.

Ruban, D.A., 2010b. Palaeoenvironmental setting (glaciations, sea-level, and plate tectonics) of Palaeozoic major biotic radiations in the marine realm. Ann. Paléontol. 96, 143–158.

Ruban, D.A., Radulovic, B.V., Radulovic, V.A., 2015. Diversity dynamics of Early and Middle Jurassic brachiopods in the Getic and Danubian tectonic units of eastern Serbia: regional versus global patterns. Palaeogeogr. Palaeoclimatol. Palaeoecol. 425, 97–108.

Sælen, G., Lunde, I.L., Porten, K.W., Braga, J.C., Dundas, S.H., Ninnemann, U.S., et al., 2016. Oyster shells as recorders of short-term oscillations of salinity and temperature during deposition of coral bioherms and reefs in the Miocene Lorca Basin, SE Spain. J. Sediment. Res. 86. Available from: http://dx.doi.org/10.2110/jsr.2016.18.

Sammarco, P.W., Horn, L., Taylor, G., Beltz, D., Nuttall, M.F., Hickerson, E.L., et al., 2016. A statistical approach to assessing relief on mesophotic banks: bank comparisons and geographic patterns. Environ. Geosci. 23. Available from: http://dx.doi.org/10.1306/eg.01121615013.

Sandoval, J., O'Dogherty, L., Guex, J., 2001a. Evolutionary rates of Jurassic ammonites in relation to sea-level fluctuations. Palaios 16, 311–335.

Sandoval, J., O'Dogherty, L., Vera, J.A., Guex, J., 2001b. Sea-level changes and ammonite faunal turnover during the Lias/Dogger transition in the western Tethys. Bull. Geol. Soc. France 173, 57–66.

Sandulli, R., Raspini, A., 2004. Regional to global correlation of Lower Cretaceous (Hauterivian-Barremian) shallow-water carbonates of the southern Appennines (Italy) and Dinarides (Montenegro), southern Tethyan margin. Sediment. Geol. 165, 117–153.

Santodomingo, N., Novak, V., Pretković, V., Marshall, N., di Martino, E., Capelli, E.L.G., et al., 2015. A diverse patch reef from turbid habitats in the Middle Miocene (east Kalimantan, Indonesia). Palaios 30. Available from: http://dx.doi.org/10.2110/palo.2013.047.

Sarg, J.F., 1988. Carbonate sequence stratigraphy, Sea-Level Changes—An Integrated Approach, 42. Society of Economic Palaeontologists and Minerologits Special Publication, pp. 155–181.

Sastri, V.V., Venkatachala, B.S., Narayanan, V., 1981. The evolution of the east coast of India. Palaeogeogr. Palaeoclimatol. Palaeoecol. 36, 23–54.

Sastry, M.V.A., Mamgain, V.D., Rao, B.R.J., 1968. Biostratigraphic zonation of Upper Cretaceous formations of Trichinopoly District, South India. Mem. Geol. Soc. India 2, 10–17.

Sastry, M.V.A., Mamgain, V.D., Rao, B.R.J., 1972. Ostracod fauna of the Ariyalur Group (Upper Cretaceous), Tiruchirapalli District, Tamil Nadu. Part I. Lithostratigraphy of the Ariyalur Group. Palaeontol. Indica 40, 1–48.

Sastry, V.V., Raju, A.T.R., Sinha, R.N., Venkatachala, B.S., 1977. Biostratigraphy and evolution of the Cauvery Basin, India. J. Geol. Soc. India 18, 355–377.

Schulze, F., Marzouk, A.M., Bassiouni, M.A.A., Kuss, J., 2004. The Late Albian-Turonian platform succession of west-central Jordan: stratigraphy and crisis. Cretaceous Res. 25, 709–737.

Sessa, J.A., Bralower, T.J., Patzkowsky, M.E., Handley, J.C., Ivany, L.C., 2012. Environmental and biological controls on the diversity and ecology of Late Cretaceous through Early Paleogene marine ecosystems in the U.S. Gulf Coastal Plain. Paleobiology 38, 218–239.

Sessa, J.A., Callapez, P.M., Dinis, P.A., Hendy, A.J.W., 2013. Paleoenvironmental and paleobio-geographical implications of a Middle Pleistocene Mollusc Assemblage from the marine terraces of Baía Das Pipas, Southwest Angola. J. Paleontol. 87. Available from: http://dx.doi.org/10.1666/12-119.

Sigwart, J.D., Carey, N., Orr, P.J., 2014. How subtle are the biases that shape the fidelity of the fossil record? A test using marine mollusks. Palaeogeogr. Palaeoclimatol. Palaeoecol. 403, 119–127.

Simberloff, D.S., 1974. Permo-Triassic extinctions: effects of area on biotic equilibrium. J. Geol. 82, 267–274.

Smith, A.B., Benson, R.B.J., 2013. Marine diversity in the geological record and its relationship to surviving bedrock area, lithofacies diversity, and original marine shelf area. Geology 41. Available from: http://dx.doi.org/10.1130/G33773.1.

Smith, A.B., Monks, N.E.A., Gale, A.S., 2006. Echinoid distribution and sequence stratigraphy in the Cenomanian (Upper Cretaceous) of southern England. Proc. Geol. Assoc. 117, 207–217.

Smith, D.L., Hayward, J.L., 2010. Bacterial decomposition of avian eggshell: a taphonomic experiment. Palaios 25. Available from: http://dx.doi.org/10.2110/palo.2009.p09-115r.

Spalletti, L.A., Queralt, I., Matheos, S.D., Colombo, F., Maggi, J., 2008. Sedimentary petrology and geochemistry of siliciclastic rocks from the Upper Jurassic Tordillo Formation (Neuquén Basin, western Argentina): implications for provenance and tectonic setting. J. South Am. Earth Sci. 25, 440–463.

Srinivasan, M.S., 1989. Recent advances in Neogene planktonic foraminiferal biostratigraphy, chemostratigraphy and paleoceanography, Northern Indian Ocean. J. Palaeontol. Soc. India, 1–18.

Steffen, M.L., 2016. Body-size trends in *Peromyscus* (Rodentia: Cricetidae) on Vancouver Island, Canada, with comments on relictual gigantism. Paleobiology 42. Available from: http://dx.doi.org/10.1017/pab.2016.14.

Stoliczka, F., 1873. Cretaceous fauna of southern India. Palaeontol. Indica 4, 66–69.

Suarez-Gonzalez, P., Quijada, I.E., Benito, M.I., Mas, R., 2015. Sedimentology of ancient coastal wetlands: insights from a Cretaceous multifaceted depositional system. J. Sediment. Res. 85. Available from: http://dx.doi.org/10.2110/jsr.2015.07.

Sugantha, T., Ramkumar, M., Balaram, V., Rai, J., Satyanarayana, D., 2015. Environmental and climatic conditions during the K-T transition in the Cauvery Basin, India: current understanding based on chemostratigraphy and implications on the KTB scenarios. In: Ramkumar, M. (Ed.), Chemostratigraphy: Concepts, Techniques and Applications. Elsevier, Amsterdam, The Netherlands, pp. 131–168.

Sundaram, R., Henderson, R.A., Ayyasami, K., Stilwell, J.D., 2001. A lithostratigraphic revision and palaeoenvironmental assessment of the Cretaceous System exposed in the onshore Cauvery Basin, southern India. Cretaceous Res. 22, 743–762.

Tanabe, K., Tsujino, Y., Okuhira, K., Misaki, A., 2015. The jaw apparatus of the Late Cretaceous heteromorph ammonoid *Pravitoceras*. J. Paleontol. 89. Available from: http://dx.doi.org/10.1017/jpa.2015.27.

Tewari, A., Hart, M.B., Watkinson, M.P., 1996. A revised lithostratigraphic classification of the Cretaceous rocks of the Trichinopoly district, Cauvery Basin, Southeast India. In: Pandey, J., Azmi, R.J., Bhandari, A., Dave, A. (Eds.), Contributions to the XV Indian Colloquium on Micropalaeontology and Stratigraphy, 789–800.

Tibert, N.E., Leckie, R.M., 2004. High-resolution estuarine sea-level cycles from the Late Cretaceous: amplitude constraints using agglutinated foraminifera. J. Foraminiferal Res. 34, 130–143.

Traini, C., Menier, D., Proust, J.-N., Sorrel, P., 2013. Controlling factors and geometry of a transgressive systems tract in a context of Ria type estuary. Mar. Geol. Available from: http://dx.doi.org/10.1016/j.margeo.2013.02.005.

Tu, T.T.N., Bocherens, H., Mariotti, A., Baudin, F., Pons, D., Broutin, J., et al., 1999. Ecological distribution of Cenomanian terrestrial plants based on $\delta^{13}C/\delta^{12}C$ ratios. Palaeogeogr. Palaeoclimatol. Palaeoecol. 145, 79–93.

Uličný, D., Jeans, C.V., 1992. Discussion on the fluctuating oceanographic conditions and glacial control across the Cenomanian-Turonian boundary. Geol. Mag. 129, 637–640.

Underwood, C.J., Goswami, A., Prasad, G.V.R., Verma, O., Flynn, J.F., 2011. Marine vertebrates from the 'Middle' Cretaceous (Early Cenomanian) of South India. J. Vertebr. Paleontol. 31, 539–552.

Ungureanu, G., Lazăr, A., Lazăr, R., Balahură, A., Ionescu, A., Micu, D., et al., 2015. Habitat mapping of Romanian Natura 2000 sites. A case study, Underwater Sulfurous Seeps, Mangalia. Ital. J. Geosci. 134. Available from: http://dx.doi.org/10.3301/IJG.2014.57.

Vail, P.R., Mitchum, R.M., Thompson, S., 1977. Seismic stratigraphy and global changes of sea-level. Part 4. Global cycles of relative changes of sea-level. In: Payton, C.E. (Ed.), Seismic Stratigraphy—Applications to Hydrocarbon Exploration, 26. Memoirs of Americn Association of Petroleum Geologists, Okla, Tulsa, USA, pp. 83–97.

Valentine, J.W., Jablonski, D., 2010. Origins of marine patterns of biodiversity: some correlates and applications. Palaeontology 53, 1203–1210.

Veizer, J., Bruckschen, P., Pawellek, F., Diener, A., Podlaha, O.G., Carden, G.A.F., et al., 1997. Oxygen isotope evolution of Phanerozoic seawater. Palaeogeogr. Palaeoclimatol. Palaeoecol. 132, 59–172.

Verma, O., 2015. Cretaceous vertebrate fauna of the Cauvery Basin, southern India: palaeodiversity and palaeobiogeographic implications. Palaeogeogr. Palaeoclimatol. Palaeoecol. 431, 53–67.

Villafaña, J.A., Rivadeneira, M.M., 2014. Rise and fall in diversity of Neogene marine vertebrates on the temperate Pacific coast of South America. Paleobiology 40. Available from: http://dx.doi.org/10.1666/13069.

Warzeski, E.R., Cunningham, K.J., Ginsburg, R.N., Anderson, J.B., Ding, Z., 1996. A Neogene mixed siliciclastic and carbonate foundation for the Quaternary carbonate shelf, Florida Keys. J. Sediment. Res. 66, 788–800.

Whitford, D.J., Allan, T.L., Andrew, A.S., Craven, S.J., Hamilton, P.J., Korsch, M.J., et al., 1996 Strontium isotope chronostratigraphy and geochemistry of the Darai limestone: Juha 1 X well, Papua New Guinea. Proceedings of III PNG Convention, 369–380.

Wignall, P.B., Bond, D.P., Sun, Y., Grasby, S.E., Beauchamp, B., Joachimski, M.M., et al., 2016. Ultra-shallow-marine anoxia in an Early Triassic shallow-marine clastic ramp (Spitsbergen) and the suppression of benthic radiation. Geol. Mag. 153. Available from: http://dx.doi.org/10.1017/S0016756815000588.

Wilson, M.E.J., 2008. Global and regional influences on equatorial shallow-marine carbonates during the Cenozoic. Palaeogeogr. Palaeoclimatol. Palaeoecol. 265, 262–274.

Wilson, M.E.J., 2015. Oligo-Miocene variability in carbonate producers and platforms of the coral triangle biodiversity hotspot: habitat mosaics and marine biodiversity. Palaios 30. Available from: http://dx.doi.org/10.2110/palo.2013.135.

Wortmann, U.G., Herrle, J.O., Weissert, H., 2004. Altered carbon cycling and coupled changes in Early Cretaceous weathering patterns: evidence from integrated carbon isotope and sandstone records of the western Tethys. Earth Planet. Sci. Lett. 220, 69–82.

Yacobucci, M.M., 2005. Multifractal and white noise evolutionary dynamics in Jurassic-Cretaceous Ammonoidea. Geology 33, 97–100.

Yadagiri, P., Ayyasami, K., 1979. A new stegosaurian dinosaur from Upper Cretaceous sediments of South India. J. Geol. Soc. India 20, 521–530.

Yadagiri, P., Ayyasami, K., 1989. A carnosaurian dinosaur from the Kallamedu Formation (Maastrichtian horizon), Tamil Nadu. In: Sastry, M.V.A., Sastry, V.V., Ramanujam, C.G.K., Kapoor, H.M., Rao, B.R.J., Satsangi, P.P., Mathur, U.B. (Eds.), Symposium on Three Decades of Development in Palaeontology and Stratigraphy in India—Precambrian to Mesozoic, 1. Geol. Soc. Ind. Spl. Publ., Bangaluru, pp. 523–528.

Yadagiri, P., Ayyasami, K., Rao, B.R.J., 1983. Cretaceous dinosaurs from southern India and their palaeogeographic significance. Rec. Geol. Surv. India 112, 51–57.

Zakharov, Y.D., Popov, A.M., Shigeta, Y., Smyshlyaeva, O.P., Sokolova, E.A., Nagendra, R., et al., 2006. New Maastrichtian oxygen and carbon isotope record: additional evidence for warm low latitudes. Geosci. J. 10, 347–367.

Zuschin, M., Ebner, C., 2015. Compositional fidelity of death assemblages from a coral reef-associated tidal-flat and shallow subtidal lagoon in the northern Red Sea. Palaios 30. Available from: http://dx.doi.org/10.2110/palo.2014.032.

Zuschin, M., Harzhauser, M., Hengst, B., Mandic, O., Roetzel, R., 2014. Long-term ecosystem stability in an Early Miocene estuary. Geology 42. Available from: http://dx.doi.org/10.1130/G34761.1.

Printed in the United States
By Bookmasters